John Older (Ed.)

Implant Bone Interface

With 141 Figures

Springer-Verlag
London Berlin Heidelberg New York
Paris Tokyo Hong Kong

M.W.J. Older MBBS BDS (Lond) FRCS (Ed)
Consultant in Orthopaedic and Traumatic Surgery,
Royal Surrey County Hospital, Guildford, Surrey, GU2 5XX, UK
Consultant Orthopaedic Surgeon, The John Charnley Hip Unit,
King Edward VII Hospital, Midhurst, West Sussex, GU29 OBL, UK
Honorary Senior Research Fellow, University of Surrey,
Guildford, Surrey, GU2 5XH, UK

ISBN-13: 978-1-4471-1813-8   e-ISBN-13: 978-1-4471-1811-4
DOI: 10.1007/978-1-4471-1811-4

British Library Cataloguing in Publication Data
Implant bone interface.
  1. Medicine. Orthopaedics. Surgery
  I. Older, John *1935–*
  617.3

Library of Congress Cataloging-in-Publication Data
Implant bone interface / John Older (ed.).
     p. cm.
  Based on the Implant Bone Interface Symposium, held in May 1989 in Midhurst, UK, sponsored by Charles F. Thackray.
  Includes index.

  1 Orthopedic implants—Complications and sequelae—Congresses. 2. Bones—Effect of implants on—Congresses. 3. Orthopedic implants—Biocompatibility—Congresses. 4. Biological interfaces—Congresses. I. Older, John, 1935–   II. Chas. F. Thackray Ltd. III. Implant Bone Interface Symposium (1989 : Midhurst, England)
  [DNLM: 1. Bone and Bones—congresses.   2. Implants, Artificial—congresses.   WE 190 I34 1989]
RD 755.5.I44   1990
617.4'710592—dc20
DNLM/DLC
for Library of Congress   90-10552
                                                                      CIP

Typeset by Goodfellow and Egan, Cambridge

Preface

As we enter the 1990s, those of us involved in implant surgery are increasingly aware of the fundamental importance of the junctional tissues between prosthesis and bone.

In May 1989 we held a symposium on the Implant Bone Interface with presentations by a distinguished international Faculty. All were authorities in their field, with a unique depth of knowledge between them, and engineers, biologists and clinicians had a rare opportunity to listen and discuss with them all the factors that influence the junctional tissues.

The meeting proved to be a constructive audit of present knowledge and this book is the result. Its purpose is to disseminate data to colleagues and to provide a yardstick for critical assessment in the future.

Midhurst   John Older
January 1990

Acknowledgements

I am deeply indebted to Charles F. Thackray for their sponsorship of the Implant Bone Interface Symposium and to the Charnley Trust for their generous financial contribution to the preparation of this book.

My grateful thanks and appreciation are also due to Bobbin Baxter for her editorial assistance; to King Edward VII Hospital, Midhurst who provided the venue; the staff, and in particular its Vice President, Lavinia, Duchess of Norfolk who graciously opened the proceedings; to Michael Phillips of Metaphor for organising the Meeting; to Robert Bohill Associates for providing such excellent sound and photographic coverage; to Ivor Williams of Palantype for his verbatim reporting; and to Tracy Bilsland for typing this manuscript.

John Older

Contents

Faculty

Professor T. Albrektsson, PhD, MD
University of Göteborg, Biomaterials Group, Department of Handicap Research, Brunnsgatan 2, S-413 12 Gothenburg, Sweden

Dr G.W. Blunn, PhD
Lecturer, Department of Biomedical Engineering, The Institute of Orthopaedics, Royal National Orthopaedic Hospital, Brockley Hill, Stanmore, Middlesex, HA7 4LP, UK

Dr I.C.C. Clarke, PhD Bio Eng, BS Mech Eng
President, Kinamed, Inc., 2192-C Anchor Court, Newbury Park, CA 91320-1603, USA

Dr A.J. Darby, MBBS, FRCPath
Consultant Histopathologist, Robert Jones and Agnes Hunt Orthopaedic Hospital, Oswestry, Shropshire, SY10 7AG, UK

Dr K. Draenert, MD
Consultant Orthopaedic Surgeon, IHSG, Gabriel Max Strasse 3, D8000 Munich, West Germany

Mr R.A. Elson, MBBChir, FRCS
Consultant Orthopaedic Surgeon, Northern General Hospital, Herries Road, Sheffield, S5 7AU, UK

Professor R.H. Fitzgerald Jr, MD
Hutzel Hospital, 4707 St Antoine Blvd, Detroit, MI 4821, USA

Professor D.W. Howie, PhD, MBBS, FRACS
Department of Orthopaedic Surgery and Trauma, Royal Adelaide Hospital, North Terrace, Adelaide, South Australia 5000, Australia

Professor L.E. Lanyon, BVSc, PhD, MRCVS
Principal, Royal Veterinary College, Royal College Street, London, NW1 OTU, UK

Dr A.J.C. Lee, BSc, PhD CEng
Director, School of Engineering, Engineering Building, University of Exeter, North Park Road, Exeter, Devon, EX4 4QF, UK

Dr L. Linder, PhD, MD
Department of Orthopaedic Surgery, Lanssjukhuset Gavle, S-801 87 Gavle, Sweden

Mr R.S.M. Ling, MA, BM (Oxon) Hon FRCS (Ed) FRCS
Consultant Orthopaedic Surgeon, Princess Elizabeth Orthopaedic Hospital,
Wonford Road, Exeter, Devon, EX2 4LE, UK
Honorary Professor of Bioengineering, School of Engineering, University of
Exeter, North Park Road, Exeter, Devon, EX4 4QF, UK

Dr A.J. Malcolm, MB ChB, MRCPath
Senior Lecturer and Honorary Consultant in Pathology, University Department of
Pathology, Royal Victoria Infirmary, Newcastle upon Tyne, NE1 4LP, UK

Mr A.W. Miles, MSc, MBES
Lecturer in Design, School of Mechanical Engineering, University of Bath,
Claverton Down, Bath, Avon, BA2 7AY, UK

Dr C. Ulrich, PhD, MD
Assistant Professor and Trauma Surgeon, Department of Traumatology, Klinik am
Eichert, 11 Oberarzt der Unfallchirugischen Klinik, 7320 Gottingen, West
Germany

Professor H.G. Willert, MD
Orthopaedische Klinik, Robert Koch Strasse 40, 3400 Gottingen, West Germany

List of Co-authors

Professor P.P. Anthony, MBBS, FRCPath
Consultant Pathologist, Royal Devon and Exeter Hospital, Exeter, Devon, UK
Professor of Clinical Histopathology, University of Exeter, North Park Road,
Exeter, Devon, EX4 4QF, UK

Dr G.C. Dracopoulos, MBBS
Orthopaedic Registrar, Department of Orthopaedic Surgery and Trauma, Royal
Adelaide Hospital, North Terrace, Adelaide, South Australia 5000, Australia

Mr G.A. Gie, MBChB, FRCS (Ed)
Senior Registrar, Princess Elizabeth Orthopaedic Hospital, Wonford Road, Exeter,
Devon, EX2 4LE, UK

Dr H. Heinrich, MD
Anaesthetist, Universitätklinic für Anaesthesiologie, Steinhovelstr. G. Ulm, West
Germany

Mr C.R. Howie, BSc, MBChB, FRCS (Ed), FRCS (Ed) Orth
Consultant Orthopaedic Surgeon, Raigmore Hospital, Inverness, Scotland, UK

Ms M. McGee, BSc
Research Officer, Department of Orthopaedic Surgery and Trauma, Royal
Adelaide Hospital, North Terrace, Adelaide, South Australia 5000, Australia

Mr R.D. Perkins, MBBS, FRCS
Senior Registrar, Princess Elizabeth Orthopaedic Hospital, Wonford Road, Exeter,
Devon, EX2 4LE, UK

Dr C.M. Steele-Scott, MBBS, FRACS
Senior Visiting Orthopaedic Surgeon, Orthopaedic Unit, Department of Surgery,
Flinders Medical Centre, Bedford Park, South Australia 5042, Australia

Miss M.E. Wait, MIST
Scientific Officer, Department of Biomedical Engineering, Institute of Orthopaedics, Royal National Orthopaedic Hospital, Brockley Hill, Stanmore, Middlesex, HA7 4LP, UK

Dr D. Waters, MBBS
Orthopaedic Registrar, Royal Adelaide Hospital, North Terrace, Adelaide, South Australia 5000, Australia

Participants

Mr D. Allan, MBChB, FRCS
Senior Orthopaedic Registrar, Western Infirmary, Dunbarton Road, Glasgow, G11 6NP, UK

Mr R. M. Atkins, MA, DM, FRCS
Consultant Senior Lecturer in Orthopaedic Surgery, Bristol Royal Infirmary, University of Bristol, Bristol, Avon, BS2 8HW, UK

Dr V. C. Bitounnis, MD (Athens)
ARC Research Fellow, University of Bristol, Bristol, Avon, BS2 8HW, UK

Dr N. Buma, MD
Head of Histology, Orthopaedic Department, Katholieke Universiteit Nijmegen Sint Radboudziekenhuis, Postbus 9101, 6500 HB Nijmegen 8, Netherlands

Ms P. Campbell, BSc
Director, Implant Retrieval Laboratory, Division of Orthopaedic Surgery, UCLA, 76–116 CHS, Los Angeles, CA 90024, USA

Mr E.H. Compton, MBBS, FRCS
Consultant Orthopaedic Surgeon, Harlow Wood Orthopaedic Hospital, Nottingham Road, Mansfield, Notts, NG18 4TH, UK

Mr J.C. D'Arcy, MBBCh, FRCS
Consultant Orthopaedic Surgeon, District General Hospital, Kings Drive, Eastbourne, East Sussex, BN21 2UD, UK

Mr N. Ditoro, BSc Eng
Lecturer in Mechanical Engineering, Chisholm Institute of Technology, 900 Dandfnong, Caulfield East, 3145 Australia

Dr A. Enderle, MD
Orthopaedic Surgeon, Orthopaedische Klinik, Robert Koch Strasse 40, 3400 Göttingen, West Germany

Dr B. Ezedinne, MD
Senior Registrar, Mohamad Hospital de Badacona, "Germans Trias i pujol", Carretera de Canyet S/N, Barcelona, Spain

Mr D.J. Ford, MBBS, FRCS
Consultant Orthopaedic Surgeon, Robert Jones and Agnes Hunt Orthopaedic Hospital, Oswestry, Shropshire, SY12 7AG, UK

Mr G.A. Gie, MBChB, FRCS (Ed)
Senior Registrar, Princess Elizabeth Orthopaedic Hospital, Exeter, Devon, EX2 4LE, UK

Professor P. Gregg, MD, FRCS
Professor of Orthopaedic Surgery, University of Leicester, University Road, Leicester, LE1 7RH, UK

Mr H.E. Griffiths, CBE, TD, MBBS, FRCS
Consultant Orthopaedic Surgeon, Southmead Hospital, Westbury-on-Trym, Bristol, Avon, BS10 5NB, UK

Dr C.J. Grobbelaar, M Med Chir(Orth), MD
Orthopaedic Surgeon, PO Box 915-1518, Faerie Glen Medical Centre 220, Faerie Glen, Pretoria 0043, South Africa

Mr D.M. Gruebel-Lee, MBBCh, FRCS
Consultant Orthopaedic Surgeon, Frimley Park Hospital, Frimley, Camberley, Surrey, GU16 5UJ, UK

Dr O. Hansen, MD
Orthopaedic Surgeon, Ekeveien 3, 1750 Halden, Norway

Mr A.J.G. Howse, MBBS, FRCS
Consultant Orthopaedic Surgeon, Central Middlesex Hospital, Park Royal, London, NW10 7NS, UK

Dr G. Isaac, BSc, MSc, PhD
Centre for Hip Surgery, Wrightington Hospital, Wigan, Lancs, WN6 9EP, UK

Mr D.P. Johnson, MBChB, FRCS
Lecturer in Orthopaedics, Bristol Royal Infirmary, University of Bristol, Bristol, Avon, BS2 8HW, UK

Mr W. MacDonald, BEng, MPhil
Senior Bio-Engineer, Johnson and Johnson Orthopaedics, Queensway, Stem Lane, New Milton, Hampshire, BH25 5NN, UK

Ms M. Magee, BSc
Research Worker, Department Orthopaedic Surgery, Royal Adelaide Hospital Adelaide, South Australia 5000, Australia

Dr K. Makela, MD
Orthopaedic Surgeon, Central Hospital of Seinajoki, Hanneksenrinne, 60220 Seinajoki, Finland

Dr D.R.W. May, BSc, MSc, PhD
Assistant Director of Bio-Medical Engineering, Institute of Orthopaedics, Royal National Orthopaedic Hospital, Brockley Hill, Stanmore, Middlesex, HA7 4LP, UK

Mr D.J.W. McMinn, MBBS, FRCS
Consultant Orthopaedic Surgeon, Dudley Road Hospital, Birmingham, B18 7QH, UK

Mr K. Nissen, MD, FRCS (Ret'd)
Prospect House, The Avenue, Sherborne, Dorset, DJ9 3AJ, UK

Mr M.D. Northmore-Ball, MA, MBBChir, FRCS
Director, Unit for Joint Reconstruction, Robert Jones and Agnes Hunt Orthopaedic Hospital, Oswestry, Shropshire, SY12 7AG, UK

Dr J. Salazar, MD
Consultant Orthopaedic Surgeon, Mohamad Hospital de Badacona, "Germans Trias i pujol", Carretera de Canyet S/N, Barcelona, Spain

Professor Dr T.J.J.H. Slooff, MD
Katholieke Universiteit Nijmegen, Sint Radboudziekenhuis, Postbus 9101, 6500 HB Nijmegen, Netherlands

Mr S.R. Smith, MBBCh, FRCS
Consultant Orthopaedic Surgeon, Newcastle General Hospital, Westgate Road, Newcastle, NE4 6BE, UK

Professor L. Solomon, MD, FRCS
Professor of Orthopaedic Surgery, Bristol Royal Infirmary, University of Bristol, Bristol, Avon, BS2 8HW, UK

Mr P.J. Turner MBChB, FRCS
Senior Orthopaedic Registrar, Northwick Park Hospital, Watford Road, Harrow, Middlesex, HA1 3UT, UK

Mr A.J. Ward, BMedSci, BMBS, FRCS
Orthopaedic Research Fellow, Bristol Royal Infirmary, University of Bristol, Bristol, Avon, BS2 8HW, UK

Mr F.A. Weber, M Med (Orth), FRCS(C)
Orthopaedic Surgeon, Suite 26, Sandton Clinic, Bryanston, South Africa

Mr R.N. Westh, FRACS
Consultant Orthopaedic Surgeon, Repatriation General Hospital, Heidelberg, Victoria, Australia

Mr J.D. Wrighton, MBBS, FRCS
Consultant Orthopaedic Surgeon, Weymouth and District Hospital, Weymouth, Dorset, UK

Introduction

M.W.J. Older

The art and science of implant arthroplasty are achieved through the combined input of orthopaedic surgeons, biologists, engineers, nurses, physiotherapists and technical staff whose goal is to give the patient with arthrosic disease a joint that is pain free and mobile. The best chance for the most effective results is to do it right the first time. If the joint fails, we have failed the patient.

One of the worrying problems today is the young person with one arthritic hip joint who is otherwise fit and well with no in-built restraints. Should the prosthesis be used with or without cement? What type of prosthesis should it be? Would an osteotomy or an arthrodesis be better for a young patient?

Whatever prosthesis or procedure is used, irrespective of the age of the patient, the longevity of the implant will depend upon fixation. But what do we mean by fixation? Absolute fixation is an unrealistic concept, but a small degree of movement between implant and bone is acceptable as long as its magnitude does not interfere with the turnover of bone.

An interface conversion film is always present on the surface of the implant. Surface energy of the implant influences this film affecting its biocompatibility, and this may be of vital importance in the early security of the interface.

There is a very wide spectrum in the junctional tissues found at the interface between implant and bone. At one extreme there may be layers of fibrous tissue and fibro-cartilage many millimetres thick, or a single layer of Uhthoff is found, and, at the other extreme, osseo-integration where no cellular layer intervenes. What controls fixation at the interface? One of the most fundamental factors must surely be the symbiosis between what the surgeon has implanted and the patient's own tissues.

Four basic facts are relevant to implant fixation:

1. The fixation is proportional to the interlock, and that is the responsibility of the surgeon.
2. Bone necrosis always occurs next to the prosthesis.
3. There is a definite relationship between bone quality and the strains and loads incurred by the patients in their daily activities.
4. Skeletal turnover occurs throughout life, an often forgotten but very important factor.

If bone cement is used, sophisticated surgical techniques allow good pressurisation into the trabecular spaces of the cancellous bone and ensure sound mechanical interlock. With an uncemented prosthesis, the surgeon tries to create circumstances that will allow patients to create their own interlock.

Bone death always occurs after an operation and is followed by a reparative phase. There should be minimal load bearing throughout the important healing process so that osseo-integration, direct contact between implant and bone without any interfering cellular layer, can take place. Some authorities believe that satisfactory long-term function can only be achieved with osseo-integration. In practical terms, however, patients often have a soft tissue layer at the interface which may persist for many years without failure. Another contention is that circumferential osseo-integration around the implant will prevent the transfer of debris, a major factor in preventing loosening which affects the longevity of implants.

When does osseo-integration occur? Some say when bone cement is used, whilst with non-cemented implants there is a divergence of opinion.

Can these questions be answered by more extensive ultrasonic and microscopic investigation? Should we even aim for osseo-integration? It is said by some to minimise late migration, cyclical movement, hydrostatic effects and the access of wear particles. Is this the true significance of osseo-integration?

There is infinite variation in the balance between the strength of the interlock and the loads applied by the surgeon and the patient. Strong interlock and minimal loading give minimal micro-movement while a weak interlock with high loading gives maximum micro-movement. Bone has a very dynamic turnover which is influenced by the joints and muscles through the loads and stresses of daily living. The resultant skeletal remodelling will affect fixation.

When does strain become excessive and affect consequent bone resorption? The trigger may be damage at the cellular layer so that the strain which exceeds the elastic limit of the junctional tissues may be critical.

There is a fine balance between healing which leads to sound mechanical interlock, and a vicious circle of bone destruction which can easily be generated if the junctional tissues are not given sufficient chance to mature

after implant surgery. This "cascade" effect can culminate in gross loosening of the implant.

Three important questions about fixation need further discussion:

1. What is the eventual fate of the micro-interlock?
2. What is the relationship between osseo-integration and micro-interlock?
3. What are the stimuli that provoke the whole process of bone resorption?

Control of the junctional tissues and the exclusion of particulate debris will be crucial factors in implant surgery during the next 10 years.

The actual design of implants can also have a considerable effect upon the junctional tissues. Designs such as the Ogee cup, which is thought to improve the pressurisation of cement, may also potentially improve the stability of the implant.

Research, as always, is vital to progress, but present economic restraints mean increasing competition for diminishing grant support. There is pressure to publish before new procedures and prostheses have been sufficiently tested, but reliable data must be made available before treatments are adopted for widespread use.

Technology has encouraged specialisation, but knowledge in depth can purchase ignorance in breadth. We must not become so engrossed in the interface that the relationship of the parts to the whole is forgotten. Sometimes even experienced physicians may treat patients on the basis of objective facts without taking their anecdotal testimony into account.

The surgeon must make the patient part of the team before, during and after the brief dramatic incident of the surgical procedure. A good total hip replacement is the product of many hands working together.

This book updates the pioneering work of John Charnley in the field of implant surgery. We have tried to answer some of the questions posed by the behaviour of the junctional tissues in order to ensure the future success of implant surgery and better care for our patients.

Part I

Histology of the Junctional Tissues

Chapter 1

An Overview of the Histology of the Principal Interface Types in Orthopaedic Surgery Today

L. Linder

I have limited this overview to stable interfaces only, with no discussion on loosening.

Biologically, bone is of soft tissue origin which can heal, after injury, with connective tissue (fibrous or cartilaginous), or with osseous tissue. When implants are inserted into the body, this healing potential is being used (Fig. 1.1).

Post-operatively, the implant and the bone are in contact, but still there are gaps and voids between the surfaces, since the contact is not uniform along the entire interface. During the healing phase, these gaps are filled with callous tissue which may mature in different directions, i.e. by bony healing, which Branemark has termed osseo-integration, where there is no fibrous lining between the implant and the bone (Branemark et al. 1977), or by soft tissue encapsulation.

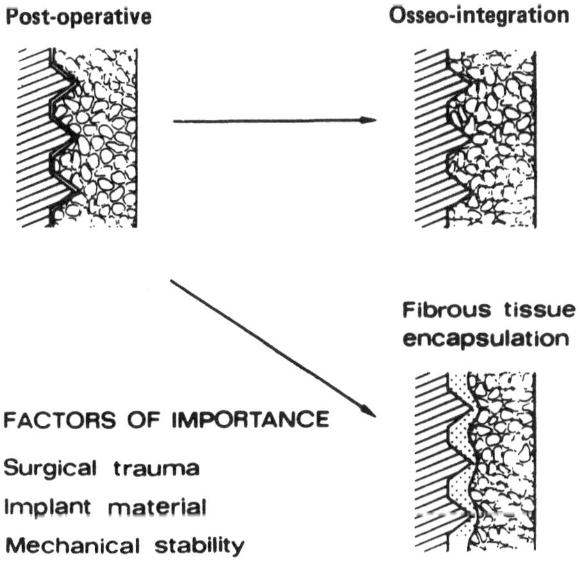

Fibrous Tissue Interface

The fibrous tissue interface is the one most commonly seen in clinical practice. The "Austin—Moore" prosthesis, an uncemented smooth-surfaced implant, is a typical example. At high magnification, a very thin radiolucent line can be seen on the X-ray around the tip of the implant. Histologically, there is a fibrous membrane of variable thickness. The collagenous tissue is parallel to the surface of the implant. The arrangement of the collagen fibres in the case of a cemented implant is principally the same. The cellular reaction adjacent to the cement surface is, however, more pronounced, especially in the superficial layer where phagocytic cells are commonly found.

From the purely mechanical point of view, these interfaces can withstand compressive and, to a lesser extent, shear loads, but are

Post-operative

Osseo-integration

Fibrous tissue encapsulation

FACTORS OF IMPORTANCE

Surgical trauma

Implant material

Mechanical stability

Fig. 1.1. Schematically, the possible pathways for tissue differentiation around implants in bone. Fibrous tissue includes fibro-cartilage.

hardly compatible with tensile loads. The implant can be pulled away from the surface of the bone and can loosen easily as this happens.

Turning to the porous-coated implants, there is obviously a connection between the bone and the implant surface which is qualitatively quite different from that which we see with a smooth-surfaced implant. When the prosthesis, a total knee replacement, was removed because of tibial loosening, the femoral component was completely stable and some of the bone remained on the retrieved component. After sectioning, the bone could be seen to be separated from the porous surface by an homogeneous layer of fibrous tissue. In polarised light, the collagen fibres were parallel to the smooth part of the surface, whereas in the porous part collagen fibres could be seen going deep into the interconnecting pores within the porous system.

This interface can also resist tensile loads since the collagen runs in different directions very much like the periodontium around the teeth or the arrangement of collagen fibres in a spinal disc.

Cartilaginous Interface

I have not so far seen cartilage at the interface in the acetabulum, although it has been described by many authors including Charnley himself (Charnley 1979), but it would seem that fibrous tissue is the predominant finding in the acetabulum. In the knee, however, I have seen it quite frequently in stable cemented knee implants on the tibial side.

The tibial component of a lateral uni-compartmental cemented prosthesis was removed for reasons other than loosening. When we sectioned that interface we saw that part of it was, in fact, fibro-cartilage. The extension of the cartilage layer could be seen. The periphery of the interface was fibrous tissue, but the central part underneath the prosthesis was cartilage. The dissolved-out cement could also be seen, with some of the barium sulphate still remaining on the surface. This fibro-cartilage must have formed after implantation.

Inside the fibrous membrane of this interface there were remnants of bone cement which were always surrounded by foreign body giant cells. However, the same kind of cement remnants could be seen in the cartilage layer, and there was no sign of any adverse tissue reaction; in fact, no cellular reaction was observed at all. This indicates that giant cells can only be found in vascularised tissues.

Cartilage, just like fibrous tissue, is capable of carrying compressive and, to a smaller extent, shear loads. Its occurrence under a tibial component must mean that the mechanical environment is suitable for cartilage to form, and probably indicates a very stable situation.

Stereophotogrammetry of these very components showed they had not migrated for over 3 years. That also supports the view that cartilage is an indicator of a stable situation.

Osseous Interface

Turning now to healing by osseous tissue, no radiolucency can be seen radiographically between the points of bone—implant contact. When we used a titanium screw of the Branemark design it was surrounded by bone histologically, and we saw mature mineralised bone without any soft tissue layer between the implant and the bone (Linder et al. 1988). At high magnification, the osteocytes were very close to the interface and there was no continuous soft tissue membrane. How often does this occur in clinical practice? From reports I have read, a number of implants have been documented to be surrounded by bone, but only to a very minor degree. In one example of a porous-coated hip prosthesis there was bony ingrowth, but the major part of the interface was made up of fibrous tissue in the space between the bone and the pores. Bone was obviously growing into and between the pores in some areas, verified by the osteocytes which were visible under magnification. This bone could not have been left at operation, so although it had to be newly ingrown bone it was confined to a very limited area of the available interface. Finally, bone cement could be seen lying directly against osseous tissue, to which I will return later.

There are a number of implant materials which have been shown to be bordered by bone (Engh et al. 1987; Lintner et al. 1986). In implant surgery we are aiming at a balance between the biomechanical requirements imposed on the interface and the tissue reaction to them. If a balance can be struck, a good prognosis will result.

How do we measure and quantify the quality of that interface? In the clinic, we have X-rays and our clinical examination but very little opportunity for histology. So how do we evaluate these different interface types? One way is to measure the change in position of a prosthesis within the bone. There are a number of centres in Sweden and internationally where stereophotogrammetry is being done. Stereophotogrammetry consists of putting small markers into the bone and the prosthetic components, taking consecutive X-rays at a 90° angle and feeding the data to a computer, which can then analyse very accurately to ± 0.1 mm any movement or change of position of the prosthesis in relation to the bone over a period of time. Migration or displacement can, of course, be analysed in other ways. When X-rays were taken 4 years apart of the same "Austin—Moore" prosthesis, for example, the tip of the prosthesis could be seen to have migrated into and almost through the lateral cortex. This migration was very obvious but it usually takes time to detect migration by serial X-rays. With stereophotogrammetry, any migration of the prosthesis can be seen more quickly.

During the first 4 or 6 months there can be a rather marked migration, often about 0.5 mm, which then subsides and perhaps even comes to a halt, but we do not yet know what happens in the very long term (Mjoberg 1986; Ryd et al. 1983). Do the components start to migrate again or are they completely stable? Can we correlate the pattern of migration with a particular histological interface?

Summary

Do the different histological types of interface give rise to different prognoses in the long term? How do we analyse the problem? What is the critical clinical level of resolution? Is it standard X-rays, stereophotogrammetry, light microscopy or electron microscopy? Where are the findings significant?

Answers to these questions can probably only be found through a cooperative effort between basic researchers and practising clinicians. It is my belief that exploration of the biological side of the interface will prove to be of profound importance for the improvement in implant longevity.

References and Further Reading

Branemark P-I, Hansson BO, Adell R, Breine U, Lindstrom J, Hallen O, Ohman A (1977) Osseo-integrated implants in the treatment of the edentulous jaw. Scand J Plast Reconstr Surg 16:1—132

Charnley J (1979) Low Friction Arthroplasty of the Hip. Springer-Verlag, Berlin Heidelberg New York

Engh CA, Bobyn JD, Glassman AH (1987) Porous-coated hip replacement. J Bone Joint Surg (Br) 69:45—55

Linder L, Carlsson A, Marsal L, Bjursten LM, Branemark P-I (1988) Clinical aspects of osseo-integration in joint replacement. J Bone Joint Surg (Br) 70:550—555

Lintner F, Zweymuller K, Brand G (1986) Tissue reactions to titanium endoprostheses. J Arthroplasty 1:183—195

Mjoberg B (1986) Loosening of the cemented hip prosthesis. Acta Orthop Scand 221:1—40

Ryd L, Boegard T, Egund N, Lindstrand A, Selvik G, Thorngren K-G (1983) Migration of the tibial component in successful uni-compartmental knee arthroplasty. Acta Orthop Scand 54:408—416

Chapter 2

Biological Factors of Importance for Bone Integration of Implanted Devices

T. Albrektsson

An overview of the experimental methods and approaches on which we base our current thoughts on the bone integrated interface is presented.

The Optical Chamber

The titanium optical chamber is simply a hollow screw. Before this implant is inserted, a short glass rod is placed inside the top part and a long glass rod inside the lower part. The screw is then inserted with these contained glass rods into the long bone of an experimental animal (Fig. 2.1).

The bone grows to incorporate this implant. It will also grow straight through the gap between the short and the long glass rods. This makes it possible to trans-illuminate the ingrown bone and its vessels in the live animal. Repeated studies of that same bone tissue compartment can be performed without having to sacrifice the animal. These animals can be kept for years and we can continue to examine the same tissue specimen. This is real unstained living bone and not a histological specimen.

Osteocyte lacunae and blood vessels containing red and white cells were magnified 500 times (with acceptable resolution) (Albrektsson 1979). We have various devices with which to measure blood-flow velocity in the bone. One is a video window technique for measuring individual capillary blood-flow velocity, another is a laser Doppler technique which allows for calculations of the segmental blood-flow rate in the bone tissue compartment (Albrektsson 1986).

What is the practical application of such a method? We have tried several approaches at our laboratory. We wanted to find out whether electrical stimulation had any significant impact on bone formation and blood vessels in the bone. Animals were subjected to 5, 20 or 50 µA using a direct current device. The method made it possible to look at the tissue before any stimulus was applied and then compare it with the tissue state during and at various times after onset of electrical stimulation. At 50 µA stimulation most animals showed a decrease of bone. However, at 5 µA, the amount of vessels improved significantly together with the amount of bone for a follow-up of 11 weeks (Buch et al. 1986) (Fig. 2.2).

Fat cells, bone tissue, soft tissue and the flow direction of vessels crossing can all be studied through the chamber. We are able to look at an animal before, and at varying times after, it has been subjected to electrical stimulation. We can compare the outcome and with various computerised methods can cal-

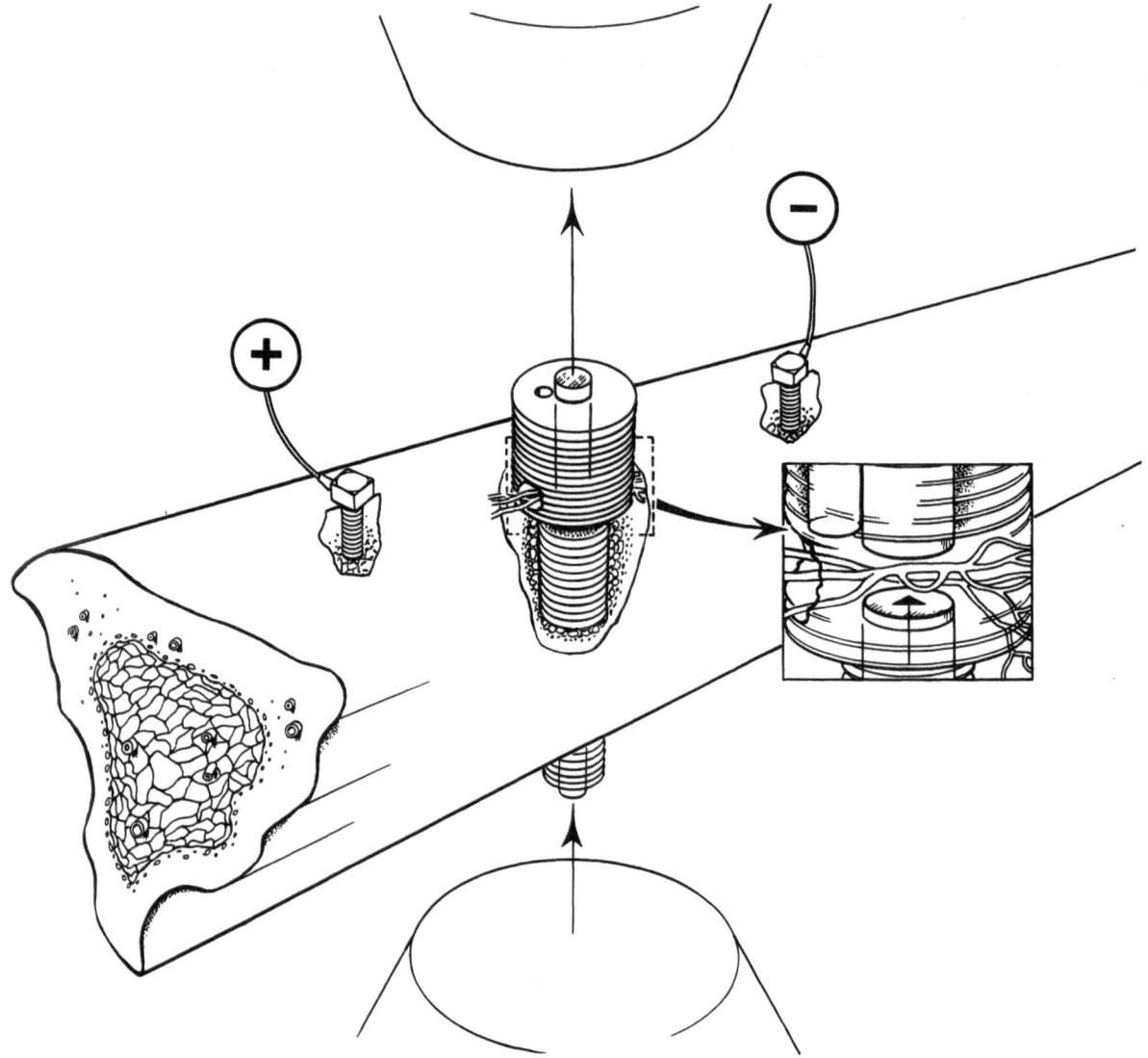

Fig. 2.1. Optical bone chamber inserted in the long bone of the rabbit. The bone accessible for vital microscopy is seen enlarged in the frame section. In this example two electrodes have been inserted to investigate the effects of applied direct currents over the chamber.

culate where the bone formation increased or decreased with the applied electrical signal.

A micro-vascular leakage after electrical stimulation can also be demonstrated by various fluorescence techniques (Nannmark et al. 1985). Is what happens in the rabbit reproducible in man? Is it clinically possible to use electrical stimulation to enhance bone growth? Here I am much more uncertain. I think that various electrical stimulating machines may have been brought into clinical practice too rapidly. But again, the tissue does react to applied electrical signals and if we could learn better control of the signals, the

method may be of a rational clinical use in the future.

The optical chamber can also be used to look at various types of trauma and their effect on bone tissue. A sensitive thermometer, i.e. a thermocouple device, was introduced into a canal parallel to the upper glass rod and measured the temperature at the site. Heat stimulus was then applied electrically. We had learnt from the literature that the critical temperature for bone was 56°C. Alkaline phosphatase is denatured at that same temperature which made it the obvious choice as the critical temperature. However,

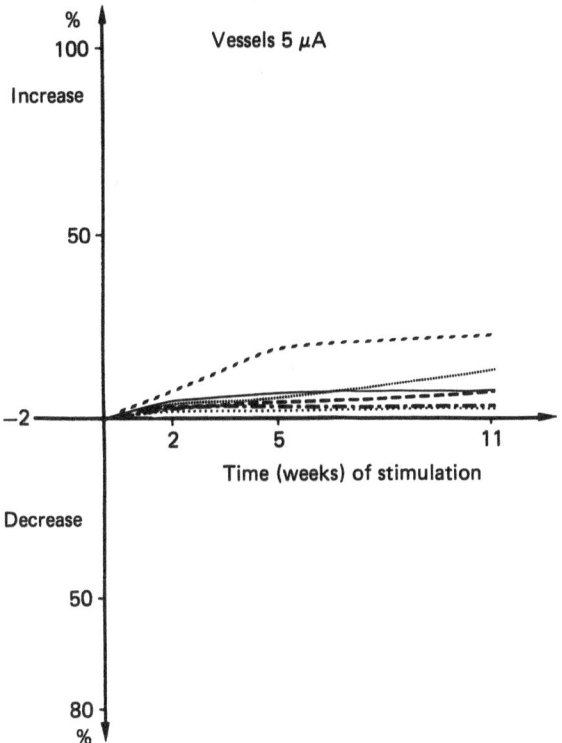

Fig. 2.2. Optical chambers were inserted in six different animals stimulated with 5 µA DC. In every case there was an increase in the amount of vessels during an 11-week stimulation period. There was a simultaneous increase in the amount of bone seen in the same optical chambers.

this fact does not rule out the possibility that temperatures lower than 56°C could be harmful to bone.

In a series of animal experiments, heating to 50°C for 1 minute showed the bone tissue to be relatively unaffected 6 days after heating. However, it takes time for bone tissue to resorb, and 3 or 4 weeks later there was always severe bone resorption after heating to 50°C for 1 minute. Of even greater interest was to follow these animals up to 1 year, when it was demonstrated that the bone tissue would never return. We are still experimenting to see if we can determine the exact time and temperature at which this irreparable state is reached. At present it is surely lower than 56°C.

Experiments have demonstrated that a temperature of 47°C applied for 1 minute seems to be the critical time—temperature relationship for bony injury as such. We concur with others that a gentle surgical technique should be used at operation so that bone and marrow tissues are allowed to heal

as such and not as low differentiated scar tissue (Eriksson et al. 1984a, 1984b; Eriksson and Albrektsson 1984).

Harvest Chamber

The harvest chamber is similar to the optical chamber, but without the glass rods. It has a canal which will penetrate the entire implant and allow for bone ingrowth. The bone that has invaded the chamber can be harvested, for instance, at 3-week intervals and this gives us several possibilities (Fig. 2.3a,b). For example, we can insert one of the implants in one of the animals and heat it to 50°C, and also have a chamber in another leg of the same animal as a control. The amount of bone formed in the previously heated site can then be compared continuously, not only immediately after the trauma but for a year or more afterwards, to see whether there will be compensation later for initial bone impairment after the heating and whether or not the healing capacity is the same as that of the control.

A micro-pump can be used that will continuously infuse a potentially bone-stimulating substance or hormone into the chamber canal. We know, because we have used radioactive tracers, that it reaches its target, and this method is currently being employed to look at various bone-stimulating agents (Albrektsson et al. 1989) (Fig. 2.4).

"Osseo-integration" of Bone Implants

Our principal aim is to unite metal and bone tissue and to observe the interfacial tissue reactions. Little is known at present about what action should be taken to control these factors. It is my belief, not based entirely on scientific knowledge, that we should aim to have a bone, not a soft tissue type of anchorage. If there is soft tissue anchorage of a metal prosthesis, and such an implant is loaded, there will be movements. This may not necessarily be critical as they are micro-move-

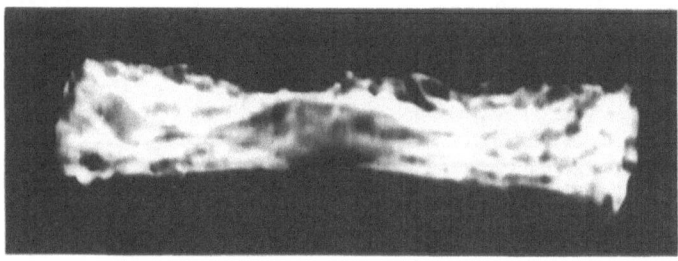

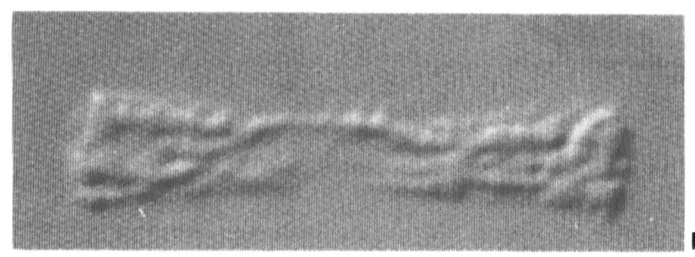

Fig. 2.3. a The harvest chamber of commercially pure titanium allows for repeated harvest of bone that invades the canal of the chamber. **b** Density curve, micro-radiogram and computerised image of the micro-radiogram. It is thus possible to quantify the amount of bone that has invaded the harvest chamber.

ments, but even in the physiological range of movement there is a constant risk of overloading the device. If overloading happens, it tends to result in macro-movements, which will lead to implant failure. However, if there are only micro-movements, as is more likely in the elderly patient, the devices, in spite of their soft tissue anchorage, may work for the lifetime of the patient.

The concept of "osseo-integration" (Fig. 2.5), i.e. direct bone to implant contact, was coined by Branemark some 10 years ago (Branemark et al. 1977). He later defined osseo-integration as a direct structural and functional contact between living bone and implant (Branemark 1985). Unfortunately, this definition raises more questions than it gives answers. What is the level of resolution for the "direct structural contact"? How do we investigate whether a "functional connection" occurs or not? Furthermore, if one single bone trabecula grows into contact with an implant and the remaining part of the interface consists of soft tissue, is that implant osseo-integrated? If not, are two or three trabeculae acceptable? It seems obvious that we have to define the amount of bone contact as well as the level of resolution we are using for the term "osseo-integration" to be verified. Furthermore, if there is osseo-integration, the

Fig. 2.4. Display of various methods used in the laboratory. *A*, The optical chamber for in vivo bone remodelling studies; *B*, The bone growth chamber for studies of the effects of electrical stimulation; *C*, The plastic plug technique for interfacial bone studies; *D*, Modified harvest chamber for the study of the effects of applied heating; *E*, A titanium implant constructed to measure the intravital bone marrow pressure.

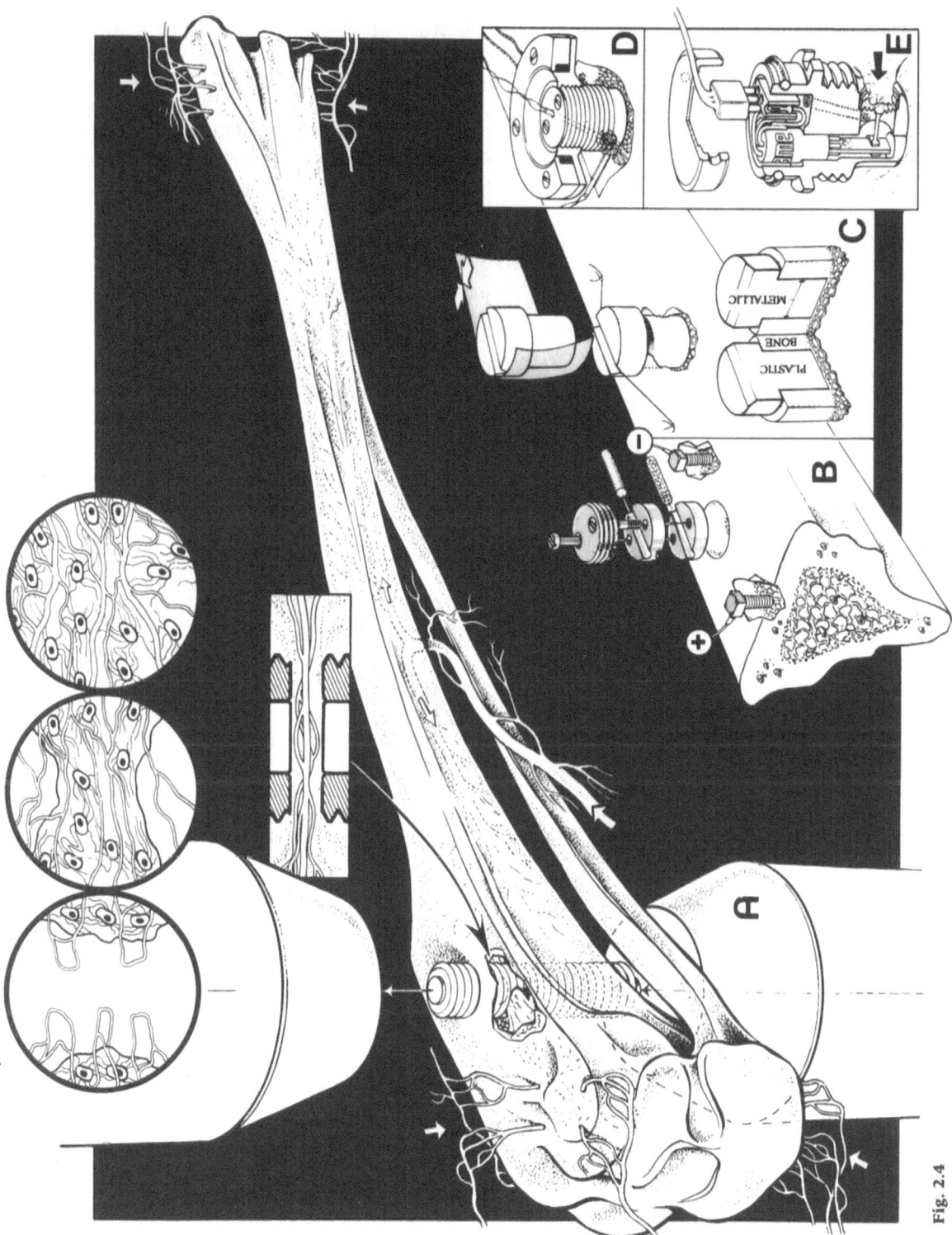

Fig. 2.4

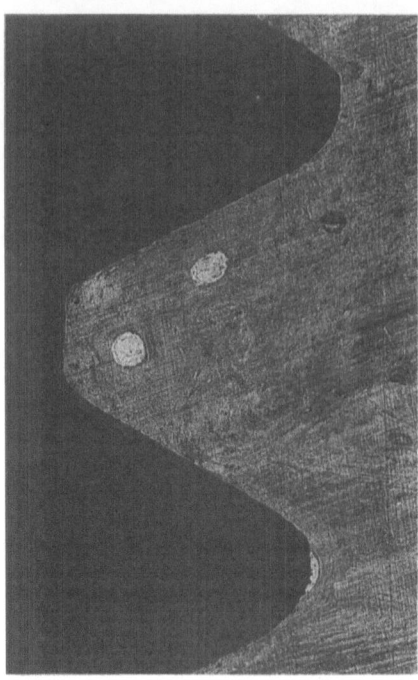

Fig. 2.5. Osseo-integrated titanium implant obtained from a postmortem specimen after 1 year of clinical function.

term as such indicates chemical—physical contact between material and bone. However, it is not surprising that the European Society of Biomaterials in their consensus conference a few years ago failed to agree upon a proper definition of the term "osseo-integration" (Williams 1987). In the absence of a generally recognised definition, let us in the future use the term osseo-integration to mean a relatively soft-tissue-free contact between implant and bone leading to a clinically stable implant. The simultaneous control of many parameters is very important to achieve clinically stable implants that do not show any microscopically observable movement when loaded.

The Nature of the Biomaterial

Chapter 6 will address some recent findings on tissue reactions to various metals, in particular a comparison between c.p. titanium and Ti-GA1-4V alloy. Of non-metallic materials, hydoxyapatite is currently a popular substance in clinical dentistry. Hydroxyapatite is a natural part of the bone and therefore quite well accepted by the tissues. However, the material is too brittle to be used in solid form and has been confined in implant dentistry as well as in orthopaedic surgery to coatings on a metal base. Such coating may become loose and several side effects may occur (Lemons et al. 1988). In fact, we lack documented long-term studies on hydroxyapatite-coated implants in dentistry as well as in orthopaedic surgery.

Implant Design

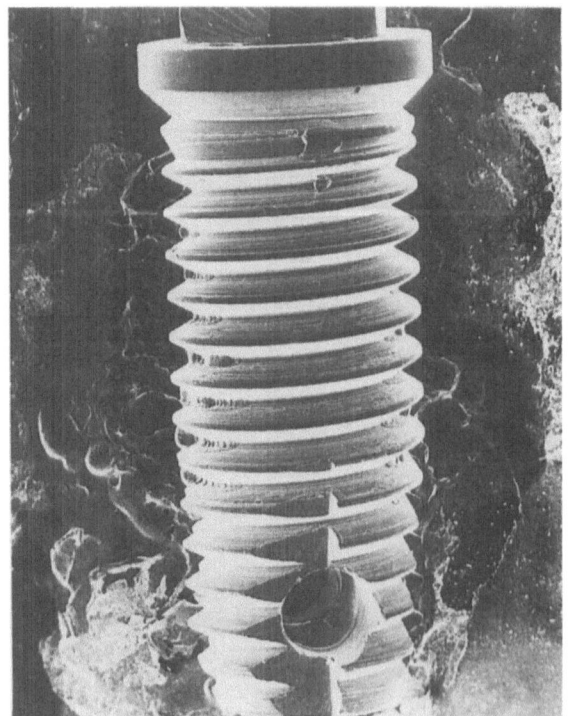

Fig. 2.6. A titanium screw design favoured by the author for use as anchorage elements for various reconstructive purposes.

Personally, I prefer to work with the threaded implants which have at least one advantage. If the implant has been inserted properly, substantial bone ingrowth is not essential for relative stability in the face of shear forces (Fig. 2.6). Three or 4 weeks must elapse with all types of porous surfaces before the bone has time to grow into the pores and stability can be achieved. Where orthopaedic implants are

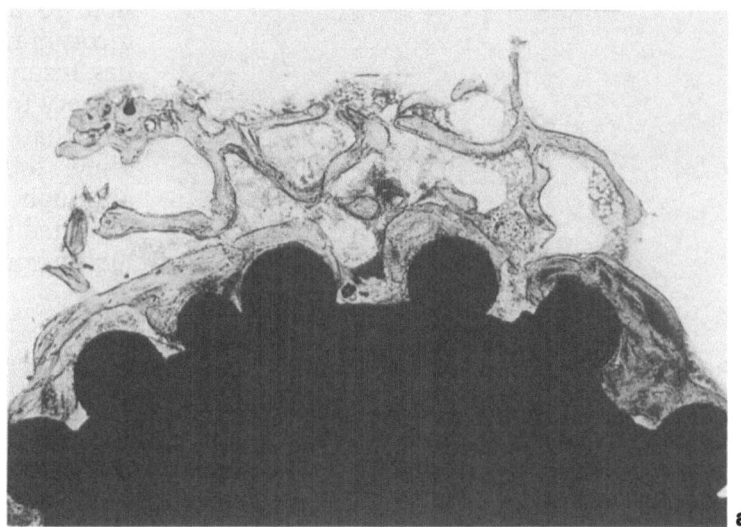

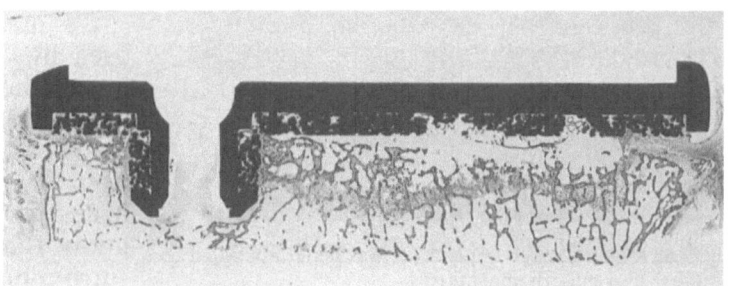

Fig. 2.7. Porous-coated implants of the hip (**a**) and knee (**b**). Generally they show only a small proportion of the implant in direct contact with the bone.

concerned it has been demonstrated that only a minor amount of the porous structures of a hip implant will become invaded by bone (Albrektsson 1990), which may depend on reasons other than the actual implant design (Fig. 2.7a,b).

Implant Surfaces

The implant surface may be a very important parameter but we lack knowledge of the "ideal" surface conditions. It could be that minor changes in the surface characteristics may affect the outcome of the implantation procedure even a long time after the actual insertion of the implant.

Biomechanically, one important parameter is the roughness of the anchoring components. A micro-rough surface is better than a smooth one (Carlsson et al. 1988) (Fig.

2.8a,b). If bone grows into the irregularities of the rough surface, an improved mechanical response is seen that can actually take much of the load (Skalak 1983). On the other hand, if irregularities are invaded by soft tissue they are of minor importance to the load bearing (Carlsson 1989). Another aspect of the surface relates to its surface energy. In vitro studies have demonstrated that high surface energy results in excellent cellular adhesion (Baier et al. 1984, 1986). Surface energy has been defined as a measure of the extent to which bonds are incomplete at the surface of a material (Hench and Ethridge 1982). One easy way of increasing the surface energy is to subject the implant to plasma cleaning or glow discharge.

We have a few ongoing studies in which we have tried to investigate the fate of such artificially enhanced implant surface energy. Are the theoretical advantages of a high surface energy beneficial in the actual bone sites of animals and man? Based on the findings of

a

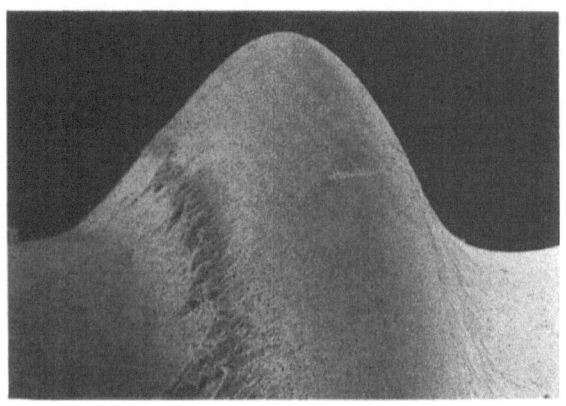

b

Fig. 2.8. A micro-rough implant (**a**) was better anchored to a bone than a smooth implant (**b**).

our studies, I am more hesitant. In the experimental part of our surface energy studies we failed to find any positive effect of such artificial enhancement of the implant energy state, but glow discharge, nevertheless, remains an interesting mode of sterilisation (Carlsson et al. 1989). The reason for the lack of effect of the high energy state may be due to the fact that most of the surface energy will be lost immediately when the implant is removed from the glow discharge machine or when it comes in contact with the tissues.

Implant Bed

We would prefer to work in a living healthy bone bed if we had the choice, but some patients have been irradiated and others have osteoporosis or related diseases. Experimentally, if one waits some time, preferably 1

year, after the irradiation procedure, even if therapeutical doses have been used, it is possible to insert an implant that will become incorporated in bone (Jacobsson 1985). This has inspired our colleagues in ENT and oral surgery to use clinical implants in spite of the fact that the patients have been irradiated. They found an acceptable success rate, although not as good as would have been expected if the tissue had not been irradiated (Jacobsson et al. 1988) (Fig. 2.9a,b).

Surgical Techniques

Minimal tissue violence is an extremely important parameter in surgical technique. We know that bone tissue is more susceptible to heat than was previously believed.

We measured the heat around the drills used by an aggressive orthopaedic surgeon, my brother, while he proceeded with internal fixation of a fracture with plate and screws. There was a temperature of 90°C 0.55 mm from the drill, but the surgeon has used these plates for 10 years and they are successful. I know that the inserted screws will become surrounded by soft tissue only. The stability of a soft-tissue-anchored screw is enough for the fractures to heal. However, if you want to insert an implant which will remain functional for the lifetime of a young human being, instead of simply treating a fracture, much more demand will be put on the interfacial structures and their capacity to carry any load put upon them. Then, the soft tissue interface, albeit successful in the fracture plate situation, will not be sufficient in the long run.

To control the surgical traumas, well-sharpened drills, low drill speeds and saline cooling must be used. I would not use a 4-mm drill if I wanted to drill a 4-mm hole but would start with a 1-mm drill and gradually enlarge this to a 4-mm size. Then it is possible to not even exceed the body temperature while drilling in bone.

When inserting a titanium screw should a gentle or firm hand be used? Only animal data are available so far. One group of implants was inserted with a relatively firm grip (average insertion torque 35 Ncm). We compared the outcome of those with other implants inserted with a gentle technique (of

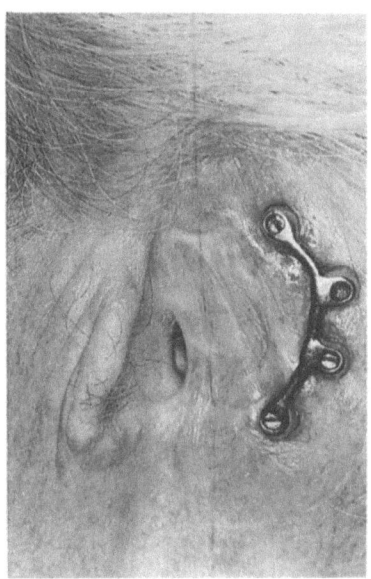

Fig. 2.9. Bone-anchored skin penetrating implants (**a**) have been inserted in previously irradiated cases for the anchorage of an auricular prosthesis (**b**). (The author is indebted to Anders Tjellstrom for permission to use this figure.)

a

b

the order of 10 Ncm). Then we removed the implants at various times. At 4 weeks, the gently inserted implants had increased in holding power and were more difficult to remove, and very much more so after 8 weeks. The ones inserted more strongly lost in holding power and continued to do so at 8 weeks.

Loading Conditions

Loading conditions are an important factor as loading too early can result in soft tissue formation. At the same time, bone must carry an adequate load and needs exercise. Branemark chose to use a two-stage surgical technique in his clinical work (Branemark et al. 1977). He inserted his implants, allowed them to heal for 3 to 6 months, then loaded them at the second stage. This is rational planning since it has been shown that prematurely loaded bone screws will move in the bone and movements stimulate the formation of soft tissue (Uhthoff 1973; Schatzker et al. 1975). However, the degree to which an implant will move is probably dependent on the bone site and it seems that rotational movements are the most harmful. We performed a series of experiments where we tried a one-stage surgical procedure and found that in the patello-

femoral joint of a rabbit, for example, a high degree of bone anchorage ensues also after primary direct loading (Rostlund et al. 1989) (Fig. 2.10). Direct loading of implants inserted in the proximal tibial region is much more uncertain however, if one wants to avoid a soft tissue anchorage. Nevertheless, it is only the implants and not necessarily the patients that have to be unloaded.

In other studies it has been demonstrated that implants that are left unloaded will show more and more bone in the interface even years after insertion onto the bone bed (Johansson and Albrektsson 1987; Tjellstrom et al. 1988; Yamanaka et al. 1989). In our recently patented knee and hip joint implants we have suggested a solution to the loading problem, although it must be pointed out that so far this has not been tried in the clinical setting.

Conclusions

We believe that direct anchorage of a bone implant may become a clinical alternative for orthopaedic implants in the future. The great clinical advantage with such directly bone-anchored cranio-facial implants has been that the losses that occur are predominantly failures during the first year of implantation.

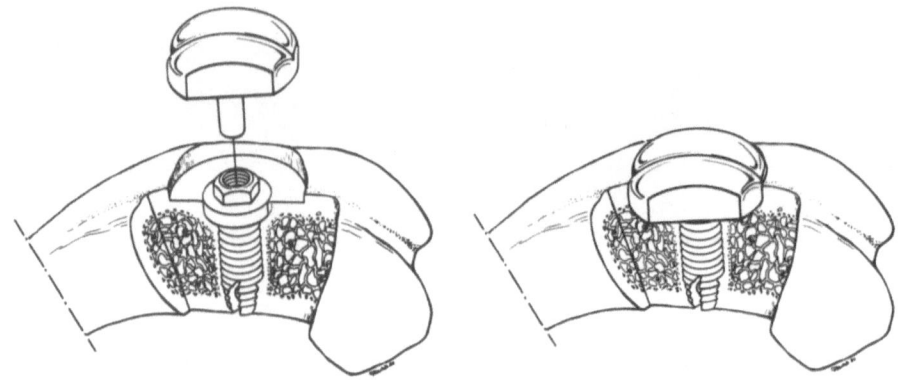

Fig. 2.10. One-stage loaded patellar prosthesis where the anchorage elements became directly bone anchored in spite of a one-stage loading procedure.

Thereafter a steady state is reached and further implant losses are unusual as the implants generally become more and more stable with increasing bone formation around them.

Acknowledgement. The author is indebted to Dr A. Tjellstrom PhD, MD, for many years of collaborative research efforts.

References and Further Reading

Albrektsson T (1979) Healing of bone grafts. In vivo studies of tissue reactions at autografting of bone in the rabbit tibia. Thesis, University of Gothenburg, Gothenburg, Sweden

Albrektsson T (1986) Implantable devices for long-term vital microscopy of bone tissue. CRC Crit Rev Biocomp 3:25—51

Albrektsson T (1990) The reactions of bone to non-cemented implants. In: Hall BK (ed) Fracture repair and regeneration. Bone: a treatise. Telford Press, Caldwell, New Jersey, 7:45—54

Albrektsson T, Eriksson AR, Jacobsson M, Kalebo P, Strid K-G, Tjellstrom A (1989) Bone repair in implant models. A review with emphasis on the Harvest Chamber for bone regeneration studies. Int J Oral Maxillofac Implants 4:45—54

Baier RE, Meyer AE, Natiella JR, Natiella RR, Carter JM (1984) Surface properties determine bio-adhesive outcomes: methods and results. J Biomed Mater Res 18:327—336

Baier RE, Natiella JR, Meyer AE, Carter JM (1986) Importance of implant surface preparation for biomaterials with different intrinsic properties. In: van Steenberghe D, Albrektsson T, Branemark P-I, Henry P, Holt R, Liden G (eds) Tissue integration in oral and maxillo facial reconstruction. Excerpta Medica Current Clin Pract Ser. 29. Elsevier, Amsterdam, pp 13—19

Branemark P-I (1985) Introduction to osseo-integration. In: Branemark P-I, Zarb G, Albrektsson T (eds) Tissue integrated prostheses. Osseo-integration in clinical dentistry.

Quintessence, Berlin Chicago Tokyo, pp 11—76

Branemark P-I, Adell R, Breine U, Lindstrom J, Hallen O, Ohman A (1977) Osseo-integrated implants in the treatment of the edentulous jaw. Experience from a ten year period. Scand J Plast Reconstr Surg 16:1—115

Buch F, Nannmark U, Albrektsson T (1986) A vital microscopic description of the effects of electrical stimulation of bone tissue. J Bioelectr 5:105—128

Carlsson LV (1989) On the development of a new concept for orthopaedic implant fixation. Thesis, University of Gothenburg, Gothenburg, Sweden

Carlsson LV, Rostlund T, Albrektsson B, Albrektsson T (1988) Removal torques for polished and rough titanium implants. Int J Oral Maxillofac Implants 3:21—24

Carlsson LV, Albrektsson T, Berman C (1989) Bone response to plasma cleaned titanium implants. Int J Oral Maxillofac Implants 4:199—204

Eriksson A, Albrektsson T (1984) The effect of heat on bone regeneration. J Oral Maxillofac Surg 42:701—711

Eriksson A, Albrektsson T, Magnusson B (1984a) Assessment of bone viability after heat trauma. A histological, histochemical and vital microscopical study in the rabbit. Scan J Plast Reconstr Surg 18:261—268

Eriksson A, Albrektsson T, Albrektsson B (1984b) Heat caused by drilling cortical bone. Temperature measured in vivo in patients and animals. Acta Orthop Scand 55:629—631

Hench LL, Ethridge EC (1982) Biomaterials — An interfacial approach. Academic Press, New York

Jacobsson M (1985) On bone behaviour after irradiation. Thesis, University of Gothenburg, Gothenburg, Sweden

Jacobsson M, Tjellstrom A, Thomsen P, Albrektsson T, Turesson I (1988) Integration of titanium implants in irradiated bone. A histological and clinical study. Ann Otol Rhinol Laryngol 97:337—340

Johansson C, Albrektsson T (1987) Integration of screw implants in the rabbit. A 1-year follow-up of removal torque of titanium implants. Int J Oral Maxillofac Implants 2:69—75

Lemons JE, Ramsay NZ, Chamoun EK (1988) Dental implant device retrievals. SFB Symposium on retrieval and analysis of surgical implants and biomaterials. 9-11 August, Snowbird, Utah

Nannmark U, Buch F, Albrektsson T (1985) Vascular reactions during electrical stimulation. Vital microscopy of the hamster cheek pouch and the rabbit tibia. Acta Orthop Scand 56:52—56

Rostlund T, Carlsson L, Albrektsson B, Albrektsson T

(1989) Osseo-integrated knee prostheses. An experimental study in rabbits. Scand J Plast Reconstr Surg 23:42—46

Schatzker J, Horne JG, Sumner Smith G (1975) The effect of movement on the holding power of screws in bone. Clin Orthop 111:257—266

Skalak R (1983) Biomechanical considerations osseo-integrated prostheses. J Prosthet Dent 49:843—852

Tjellstrom A, Jacobsson M, Albrektsson T (1988) Removal torque of osseo-integrated craniofacial implants. A clinical study. Int J Oral Maxillofac Implants 3:287—289

Uhthoff HK (1973) Mechanical factors influencing the holding power of screws in compact bone. J Bone Joint Surg (Br) 53:633—642

Williams DF (1987) Definitions in biomaterials. In: Williams DF (ed) Progress in biomedical engineering. Elsevier, Amsterdam Oxford New York Tokyo, pp 1—72

Yamanaka E, Tjellstrom A, Jacobsson M, Albrektsson T (1989) Long term observations on removal torque of directly bone-anchored implants. Int J Oral Maxillofac Implants (In press)

Discussion

The Chairman - **Mr Older**

The Panel - **Professor Albrektsson**
            - **Dr Linder**

Mr Older: I thought that some of the comments towards the end of Professor Albrektsson's presentation were particularly significant where he showed the range from experimental work to patient care, and the importance in hip surgery today of holding back our patients from doing too much too quickly.

I recall Sir John Charnley doing a ward round 10 years ago. If somebody had had a bilateral simultaneous hip replacement and everything had gone beautifully, he would have the patient walking down the corridor as quickly as they could with one stick in the right hand. He would turn round and say, "John, it's fantastic — two hips, 14 days, and they are walking with one stick!" These days I now hold back patients and I think Sir John also realised the need for this just before he died.

Professor Solomon: What evidence is there that what happens in the animal happens in the human being, especially those of 60 and 70 years plus? That, surely, is the crucial question in the last resort?

Professor Albrektsson: Essentially, there is only one honest response to your question. In

animal research we can only determine those things that are very harmful to animals, and those we should avoid. Every time you enter the clinical theatre it is a new experimental situation. We must accept that. We do not know what will happen. It is wonderful in rabbits for 1 to 5 years. There is only one way to find out if it is good for the patients and that is by thorough clinical investigation. I can give you no such data from the orthopaedic field in my experience.

Dr Linder: I have done a clinical experimental series in which I have implanted the screws which Branemark uses in the jaw into the tibia of patients awaiting knee surgery. The amount of bone that we saw in Professor Albrektsson's slides is not present in the human situation. We do have bony contact osseo-integration. Osseo-integration has become a matter of definition. I prefer to say that it is the contact between living bone and an implant without a continuous soft tissue layer and without any signs of inflammation — no quantitative aspect at all.

Mr Smith: What about the dynamics of the fibrous interface? Does this interface constantly undergo maturation? Can it become bold? How long does it take?

Professor Albrektsson: That is a difficult question. I am a cell biologist and I would like to wait 3 months of a cellular cycle before I would expect any substantial amount of bone in the interface.

As a purist, I prefer a bone interface, but I do recognise that there are different types of soft tissue interfaces and some look very similar in appearance to the periodontal ligament. Maybe such a fibrous interface is as good as bone integration.

Dr Linder: Perhaps, but not quite. An animal study in dogs showed the pull-out strength of bone grown into pores with an arrangement of fibrous tissue. There is a fundamental difference between that kind of interface and the usual fibrous tissue lining that we are seeing around solid implants. We must do standardised experiments in order to comment on such an interface. It is a potentially very interesting interface.

In the clinical situation, if we aim for

ingrowth into pores and get fibrous tissue ingrowth instead, it may be better than the usual fibrous lining that we see around cemented implants. But that remains to be seen.

Mr Compton: I noted the temperature during the drilling experiments. In one of my own operations, I noted that there was not nearly so much difference when going through cancellous bone, only of the order of 5°C rise per mm from the saw cut, compared with the 30°C to 40°C that I found when trying to do that in the cortical bone. Irrigation with saline helped with the cortical bone but did not make any real difference with cancellous bone. If we are talking about joint replacement surgery we should be worrying much more about what we do in cancellous bone with the surface replacements than what we do with cortical bone in the drilling experiment, which is a transient situation. What we do in the cancellous interface for implant surgery is perhaps more important. But I found only perhaps 5°C difference with or without irrigation.

Professor Akbrektsson: I agree that the elevation of the temperature while drilling is much greater in the cortical bone. The relevance of that relates to where we want to insert our implants. We are presently working, at a very immature stage I must admit, with orthopaedic implants when we have tried to have them bone integrated. Those implants do not exactly resemble the conventional type of non-cemented implants used today, so they are dependent on a cortical bone anchorage. I believe that is important, and the more conventional types of implants may be anchored in cancellous bone and less prone to overheating while drilling in the bone. We measured in the cancellous bone and found much lower temperatures compared with the cortical bone. If you want to drill trans-cortically you do have to pass one or two cortical layers. It is the thickness of the cortex that is important. In the literature it says that very insignificant temperatures have been recorded even when drilling in cortical bone. Why is that? I believe the reason is that people have been working with animals with a cortical thickness of 1.0 mm or 1.5 mm, so that they are quickly through the bone and

the temperature does not rise. But a man can have 6 or 7 mm in the femoral region, and that difference is much more significant, of course, when going straight through the cortex. There is an average of 10°C higher temperature when drilling in the remote cortex on the other side where the cooling agents are less prone to reach their target than you find if you are doing a straight-through-the-bone operation. I agree totally with you that cancellous bone is less prone to heating because of that.

Mr Gruebel-Lee: Loading is a critical subject for clinicians. When you are putting in a threaded implant, is that a special case because of the torque you apply? Are you applying some degree of physiological loading which is critical for bony ingrowth and a good result?

Dr Linder: As a practising orthopaedic surgeon I am unable to give you an answer to that question. I think it is preferable to have the patients guard their weight bearing for some months after surgery, but in rheumatoid cases, for example, we just have to let them load. I do not know of any study which shows the difference radiologically. I know that Amstutz has hinted that in one group of patients who had loaded the implants from the start, he saw wider membranes or radiolucencies than in the other group who had been on crutches. That may just lend credit to having the patient on crutches, but I do not know the degree of loading.

Professor Albrektsson: In theory, the ideal bone biology would be to have a gradual increase of load. The problem, however, is to define the degree of load to the patient.

Mr Gruebel-Lee: There is a profound difference in the recovery time and the results of soft tissues. I believe that we have to trust bone. I suspect that we should tell the patient to respond to his own bone and load in that way.

Mr Older: I have done a bilateral simultaneous hip replacement in a patient under 70 who is fit and well, with some justification, being pleased with the results so far. Mr Ling said quite definitely that he felt that for somebody in his late fifties perhaps this is no longer justified. That has made me think about the whole rationale of treating people with bilateral hips.

Mr Ling: I used to do simultaneous bilateral arthroplasties but do so only very rarely now. There is no way to restrain the loading of the device and you may get bilateral ectopic bone which is a catastrophe.

What I feel now is that early loading does not matter in the femur very much but perhaps it does in the socket. Other evidence shows that the very best sockets in the long term are in arthrodesis converted to arthroplasty. Those patients do not have the muscles to generate high loads for quite a long time after the operation, so perhaps it is important in the socket.

Professor Albrektsson: The problem in discussion is that we tend to talk about one parameter and then another. We all agree that loading is very significant, but loading alone is not the problem. We have to look at many parameters, and perhaps a few others that we still do not yet know, at more or less the same time and control them all. Using current orthopaedic devices it is very difficult to control, for example, the surgical trauma. How can we do that? Let me propose, for the purpose of discussion, that that kills the bone too much for it to heal, so that if you unload it it does not matter that much.

Professor Howie: Loading tells us what is happening at the interfaces in the first year. I should like the presenters to comment on our findings, which have been that the response in a cemented hip in the femur in the first year is at a much slower rate than one sees in animals. Dead bone is still present a year after the insertion and the remodelling is still going on. For how long does one unload a hip? In the necrotic bone, which is weak, the repair process is continuous for a year or two after insertion, and I find it difficult to imagine how I can unload for that length of time.

Dr Linder: I have done two retrieval studies of stable femoral components, and there is no doubt that even after more than 2 years necrotic islands of bone will still be seen adjacent to the cement. The entire interface consists of dead bone immediately after surgery and there is a gradual exchange of dead for living tissue. The problem is to maintain the mechanical integrity of the interface during this time. Any suggestions of unloading are only arbitrary. Is 2 months enough, or should it be 3 months or 4 months? Nobody knows. The movements produced by the loading may be a more critical factor than the loading itself.

Professor Albrektsson: I totally agree, we cannot unload a patient's limb for months and months. The only one to benefit from that would be the histologist. I do not think, therefore, that is a point for discussion. What we can do, without much harm to the patient, is to unload the implant.

Chapter 3

The Reaction of Bone to Bone Cement in Animals and Humans

L. Linder

John Charnley revolutionised orthopaedic surgery and it is impossible to discuss the reaction to bone cement without this historical perspective, because our view has changed over the years. In the 1960s and early 1970s, what we discussed was whether a 2-mm radiolucency or a 1-mm radiolucency was acceptable and whether it was the normal response of bone to bone cement. If we looked at the interfaces histologically in those years, we invariably found a thick fibrous membrane between the bone and the bone cement. Why was the fibrous tissue there? Was it because of the bone cement as such, or the wear of the polyethylene socket? Was it due to metal corrosion, or the fragmentation of the bone cement surface? All these phenomena could be traced in the membrane, which was very bewildering (Mirra et al. 1976).

If these membranes are looked at from the enzyme histochemical point of view, an ominous sign in a membrane which is supposed to be stable is the presence of a large amount of acid phosphatase. At high resolution, an osteoclast containing acid phosphatase can be seen chewing away on living and vital bone, not dead necrotic bone. That, coupled with the presence of alkaline phosphatase which indicates osteoblastic activity, shows that there is a high rate of bone turnover at the

border of the soft tissue and the bone. One may wonder what is the stimulus for this bone turnover.

Initially, a disturbing difference appeared between findings in animals and in humans which led people to believe that there was a species difference between the response to bone cement in animals and in humans. In animals the response was quite easy to reproduce many times over, and in different species a direct bone—bone cement contact could be produced. Various factors influence the morphology of the interface such as the biocompatibility of the implant itself, the mechanical loading of that interface during implantation and the tissue condition after surgery (or even pre-operatively). The experimental situation is much simpler because we can isolate whatever we want to look at and try to make all other factors as equal as possible.

An experimental study was undertaken with a cylindrical implant of polymerised bone cement covered with a very thin coat of titanium (Linder and Ivarsson 1986). I removed the titanium in places around the interface so that I got a mosaic surface consisting of bone cement and titanium which thus were exposed to the same kind of tissue and mechanical conditions (Fig. 3.1). At sacri-

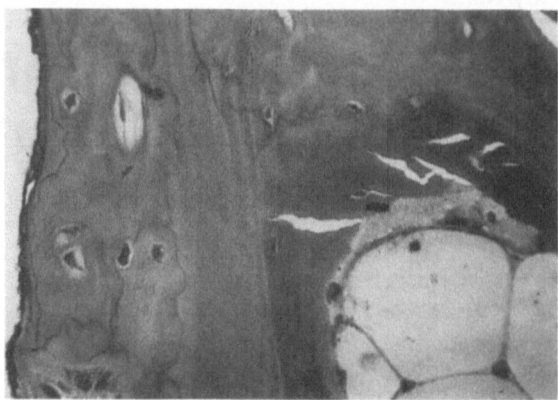

Fig. 3.3. Example of bone in direct contact with the titanium coat (*black line*), indicating that optimal conditions for bony healing have been present around this implant. (Methylene blue, ×500)

Fig. 3.1. Photograph of an implant of polymerised bone cement covered with a coat of pure titanium which in certain areas has been removed to create a mosaic-like implant surface. Since the tissue reaction to titanium is well documented, it serves as an internal control when evaluating the reaction to the polymer.

fice this implant was removed together with the surrounding bone and it was all embedded in plastic (Fig. 3.2).

During the embedding process the bone cement was dissolved, leaving the ring of bone surrounding the implant and the titanium coat on the inner side of that ring. When the ring was divided, the bone and surround-

Fig. 3.2. The mosaic plug and surrounding bone are embedded in plastic. During the embedding process the bone cement is dissolved, leaving the titanium coat on the surface of the interface tissue. The interface can now be sectioned on a microtome in routine fashion.

ing tissues were covered with the titanium coat and the places where the bone cement had been in direct contact with the tissue were visible as windows in the titanium.

A direct contact between the bone and the titanium coat could be seen, so the titanium part of the plug was osseo-integrated (Fig. 3.3). That should indicate that the implant had had optimal conditions for bony healing, which would be true even for the places where the bone cement was interfacing with the tissue. In those areas, no continuous soft tissue membrane could be seen at high magnification although there were some vascular channels. That suggests to me that the normal response to polymerised cement is a direct contact between bone and cement.

Why is it that the implants used in clinical practice are still bordered by fibrous tissue? Another case gives some food for thought in that respect, because four very distinctly different reactions to this particular implant were observed (Fig. 3.4). Initially, the usual radiolucency in the acetabular region was observed; then the gross bone resorption in the calcar region followed by the very close contact between the cement and the bone in the femur were seen; finally, there was a scalloping effect in the lateral projection where there was a very marked bony erosion. The stem of the implant was loose within the cement.

I participated in the revision operation and took biopsies from several places. Very thick membrane could be seen in the acetabulum

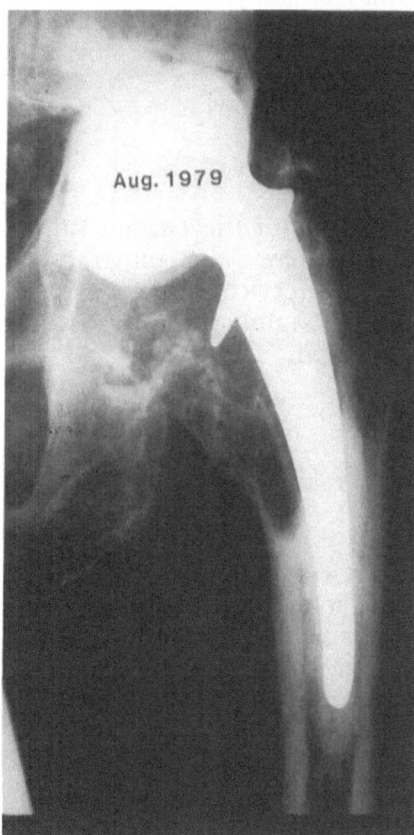

Fig. 3.4. A McKee—Farrar hip prosthesis with seemingly variable tissue reactions adjacent to the cement: a radiolucent line in the acetabulum, osteolysis in the upper femur and a close bone—cement contact in the distal portion.

resting on viable bone. Part of the membrane was completely necrotic. The basal part was vital adjacent to viable bone. Osteoclasts could be seen together with new bone formation. This represents one type of reaction. In the scalloped area, there was a fibrous membrane and a cellular reaction in the basal part together with many wear particles.

It was interesting to note that in the part where there was a close cement—bone contact radiographically, we found an interface without any soft tissue membrane histologically. Under the electron microscope, there were no signs of a continuous cellular layer between the bone and the cement (Linder and Hansson 1983). In some areas there was a cement—soft tissue contact, and here we could see flattened giant cells which Charnley has described as quite common around bone cement (Charnley 1979). In summary there were four main types of reaction in this

patient. I suggest that the reaction with a direct bone—cement contact must be the basic response, which leads me to believe that the tissue responses to bone cement in animals can be reproduced even in human beings.

We have made further enzyme histochemical studies of stable interfaces in the femur (Linder and Carlsson 1986). Unfortunately, the cement had to be removed before the bone could be sectioned, but we were lucky in some cases to see remnants of bone cement, for example a sphere of bone cement embedded in bone. This sphere could not very well have been pressed in during operation. I think that the bone had regenerated around the sphere and there was an actual bony ingrowth into the irregularities of the bone cement surface.

In another case the cement had been dissolved out. The contour of the cement and apparently viable bone in direct contact with it could be seen although there were areas of poorly mineralised bone which are quite commonly seen around bone cement.

As before, there were foreign body giant cells, but I have no idea about their function or whether they are harmful or not. These interfaces had been fully weight bearing for many years.

As a working hypothesis these findings made me believe that the normal response of bone to bone cement is a direct contact between osseous tissue and bone. But what happens in cancellous bone? Nobody has ever described a direct contact there. Is there a difference between the reaction in cancellous bone and in cortical bone? Radiologically a rather thin, but nevertheless distinct, fibrous membrane can still be seen histologically in the seemingly perfect interface in the acetabulum. It cannot be seen radiographically, however, because of over-projections. In other places in the same patient there was mostly osteoid but nevertheless osseous tissue in direct contact with the bone, and no continuous membrane.

Another example showed the same result. In the tibial component of a failed ankle prosthesis which had to be converted to an arthrodesis (Fig. 3.5), I studied the interface histologically, and the result is shown in Fig. 3.6. Seemingly normal marrow adjacent to the dissolved-out cement could be seen as

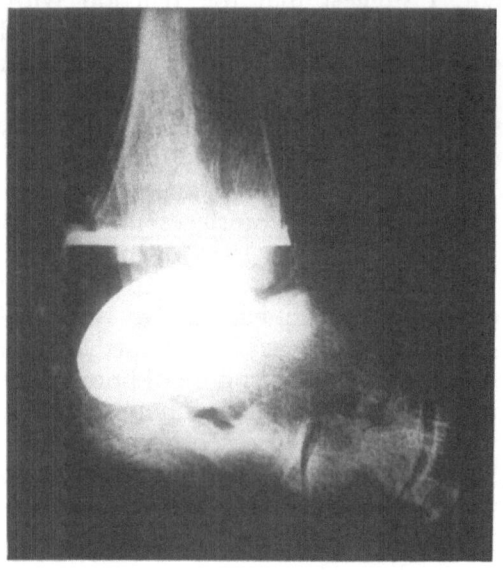

◄ **Fig. 3.5.** A cemented ankle prosthesis with mechanical loosening of the talar component but with an entirely stable tibial component.

well as remnants of the barium sulphate. In other places there was mineralised bone apparently in direct contact with the cement surface, and certainly no soft tissue membrane in between.

Summary

I believe that contact is possible, even in the load-bearing state, between bone cement and osseous tissue in the human, but the problem still remains as to how this can be achieved on a regular basis.

References and Further Reading

Charnley J (1979) Low friction arthroplasty of the hip. Springer-Verlag, Berlin Heidelberg New York

Linder L, Hansson H-A (1983) Ultra-structural aspects of the interface between bone and cement in man. J Bone Joint Surg (Br) 65:646—649

Linder L, Carlsson A (1986) The bone—cement interface in hip arthroplasty. Acta Orthop Scand 57:495—500

Linder L, Ivarsson B (1986) Evaluation of the bio-compatibility of polymers implanted into bone using titanium mosaic on bone cement. Biomaterials 7:17—19

Mirra JM, Amstutz HC, Matos M, Gold R (1976) The pathology of the joint tissues and its clinical relevance in prosthesis failure. Clin Orthop 117:221—240

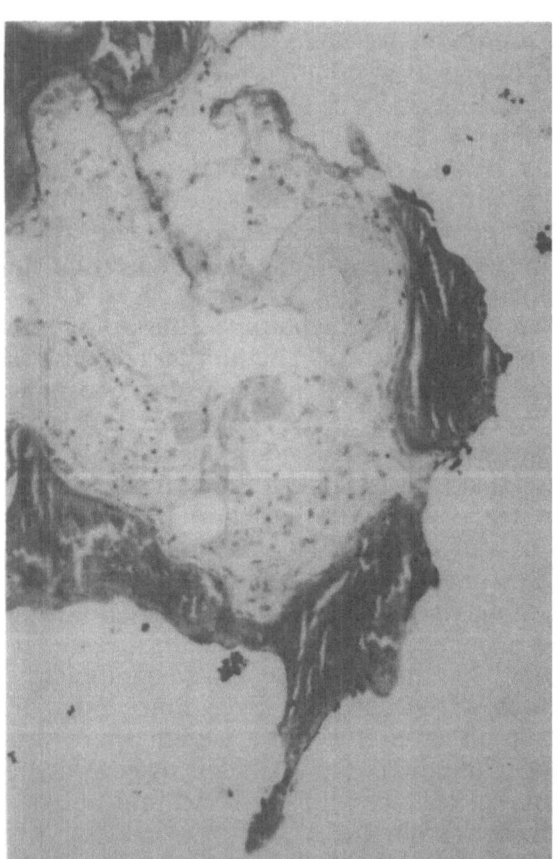

Fig. 3.6. Histological section of the bone—cement interface in the tibia of the same ankle prosthesis as in Fig. 3.5. The radio-opaque black particles of the dissolved-out cement are seen. There is apparently healthy tissue present with direct contact between cement and viable bone and osteoid. Such contact is not exclusive to cortical bone. (Goldner, ×250)

Chapter 4

New Observations on the Bone Cement—Bone Interface

K. Draenert

The clinical and histological data on the bone cement—bone interface can be investigated by animal experiments, in vitro models or studies of retrieved postmortem specimens of replaced joints (Draenert 1981). The in vitro model has shown that there is no gap between the metal rod and the bone cement, but there is one between bone cement and the bone. Animal experiments have shown that this gap was filled initially by a haematoma. Later, regenerating vessels appeared, followed by new bone formation. Four weeks after the implantation, the gap was bridged by newly formed bone. The first framework was of concentrically reinforced woven bone which was converted to compact bone by mature lamellae. This adaptation and the remodelling was completed in 2 years (Draenert 1981).

In order to study the different deformation patterns of bone under load around the implant, it is necessary to reconstruct a three-dimensional model using sequential cross sections of retrieved specimens. Histologically, this model will also enable the principles of anchorage to be identified.

The stereogrammatical investigations of the PCA knee showed very rapid migration of the components (Ryd 1986). This was confirmed histologically by thick fibrous tissue in between the bony baseplate and the coating (Haddad 1986; Draenert 1988).

A second widespread anchoring principle is the press-fit philosophy. A Zweymuller prosthesis was examined more than 3 years after implantation. The three-dimensional examination elucidated the principle of transmission of forces. Bone atrophy took place in the proximal part of the femur contrasting with the thickening of its diaphysis around the distal third of the component. At the interface there was a thick fibrous tissue layer at the level of the calcar femoris, decreasing in thickness distally to the tip of the prosthesis. Direct bone contacts were revealed around the distal half of the components. These bony contacts, free of fibrous tissue, first appeared antero-laterally, and later around the whole circumference of the tip of the prosthesis indicating a distal press-fit anchorage.

A clear conclusion can be drawn from the cemented self-locking prosthesis which is the principle of press-fit. The three-dimensional reconstruction of cadaveric specimens of replaced joints up to 8-years after the operation revealed a complete destruction of the cement sheath and a milling of the material between bone and the metal (Draenert 1986) (Fig. 4.1 opposite).

The picture which emerged for a standard prosthesis from the same manufacturer was quite different. A well-cemented component retrieved 7 years after the implantation revealed an intact cement sheath integrated to bone over the proximal two-thirds of its length. A thin fibrous tissue layer was formed distally, indicating an incomplete cementing technique around the tip of the prosthesis. The remodelling of the femur appeared physiological, with a deep penetration of compact bone into the cement sheath without the interposition of any fibrous tissue (Fig. 4.2 opposite). High-density polyethylene (HDPE) debris was found in the osteolytic formation. The three-dimensional reconstruction showed a physiological proximal anchorage of the component revealing the preservation of the

cancellous bone framework where it was stiffened by bone cement.

References and Further Reading

Draenert K (1981) Histomorphology of the bone-to-cement interface: remodelling of the cortex and revascularisation of the medullary canal in animal experiments. In: Salvati EA (ed) The hip proceedings of the ninth open scientific meeting of the Hip Society. The John Charnley award paper. CV Mosby, Saint Louis, Missouri, pp 77—110

Draenert K (1986) Histomorphologische Befunde zur gedaempften und ungedaempften Krafteinleitung in das knocherne Lager. Vereinigung Nordwestdeutscher Orthopaeden. 36. Jahrestagung 15—18 Juni, Hannover

Draenert K (1988) Forschung und Fortbildung in der Chirurgie des Bewegungs-apparates. Zur Praxis der Zementverankerung. Art and Science, Munich

Haddad RJ (1986) Histological observations of tissue ingrowth in retrieved human joint compartments. In: Draenert K (ed) Die Implantatverankerung Symposium in Orthopaedie und Chirurgie des Bewegungsapparates. Art and Science, Munich, p 30

Ryd L (1986) Micromotion in knee arthroplasty. A roentgen stereophotogrammetric analysis of tibial component fixation. Acta Orthop Scan (Suppl) 220:1—80

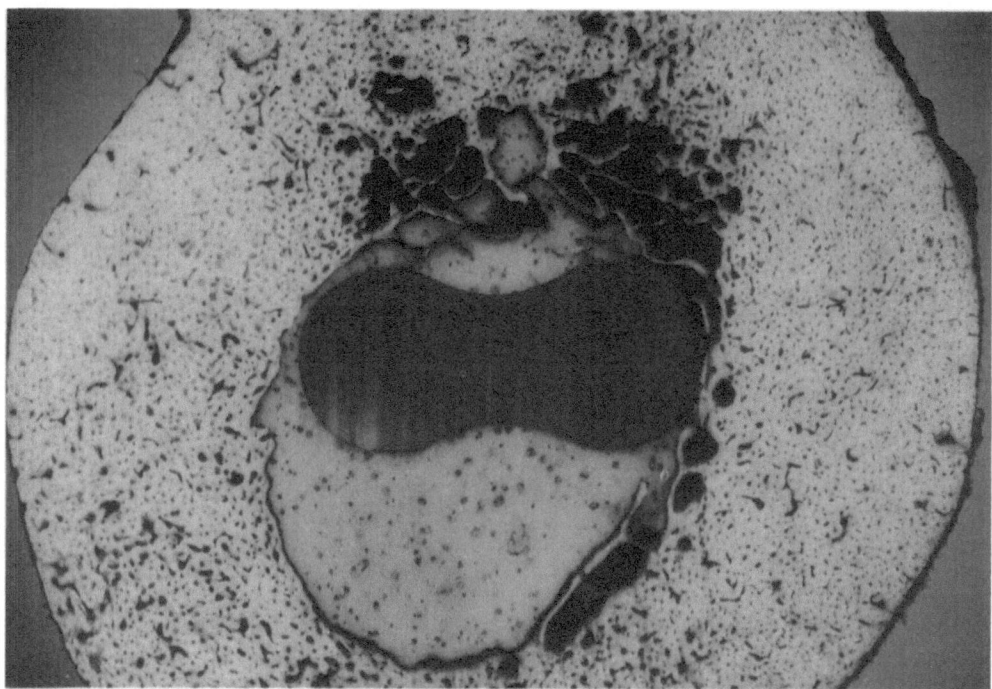

Fig. 4.1. Cross section of a cemented press-fit component at the mid-shaft level 2 years after the operation. The breakdown of bone cement is clearly pronounced and a fibrous tissue in the interface is revealed.

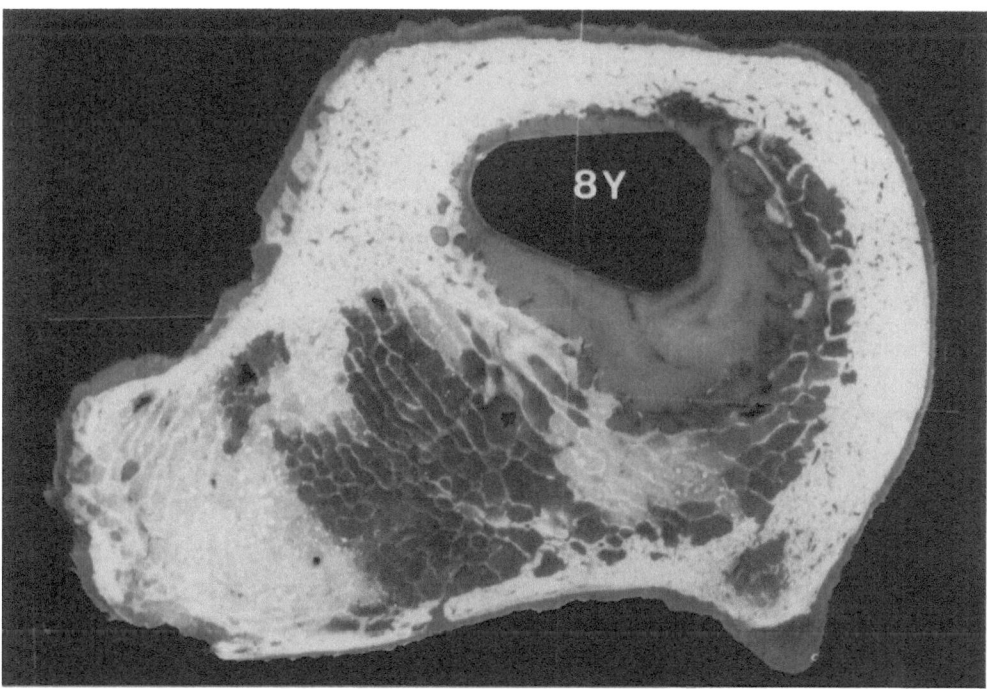

Fig. 4.2. Cross section of a standard component 8 years after the operation; the cement is still intact and a fibrous tissue-free contact presents the osseo-integration of the implant.

Chapter 5

Morphology of Implant—Bone Interface in Cemented and Non-cemented Endoprostheses

H.G. Willert

Very little knowledge was available in 1965 when we started to investigate the morphology of the PMMA bone cement—bone interface in patients. In the early 1970s we could already achieve a reasonable interlock between the bone and the cement without extra pressurisation (Fig. 5.1). It was the aim of further development of implantation techniques to improve not only the mechanical qualities of the cement cuffs but also the intrusion of the cement into the marrow spaces in order to achieve as close a contact to cortical bone as possible not only in exceptional cases but regularly.

Despite widespread research and development over a 20-year period of clinical practice the application of PMMA bone cement still presents many unanswered questions concerning its biocompatibility and stability, the nature of cellular contact of bone and other tissues to cement, the role of the soft tissue as shock absorber, the role of interlocking and the influence of load distribution.

We do considerable damage when we prepare the bed for our implant, destroying the marrow, parts of spongy and cortical bone as well as significant parts of the endosteal blood vessels. Inserting the bone cement plug and implant may introduce toxic components but

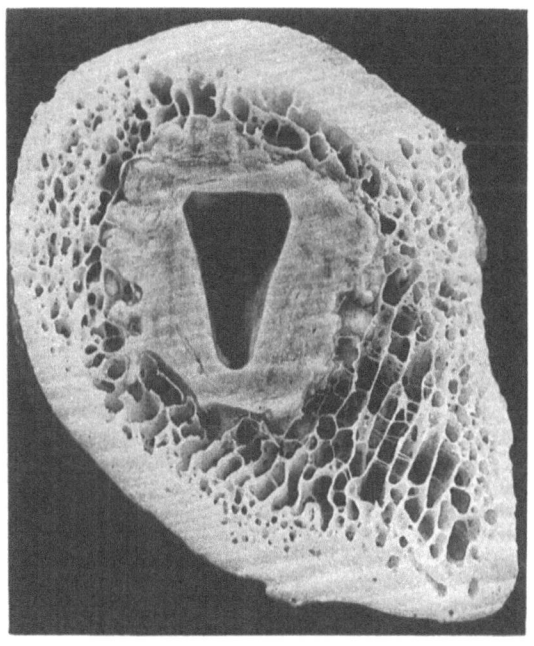

Fig. 5.1. Interlock of PMMA bone cement with bone. State of the art in the 1970s of a regular cementing technique. A macerated cross section through a proximal femur after implantation of a femoral component with PMMA bone cement demonstrates favourable features of the implant anchorage: (a) the cement dough was pressed sufficiently deep into the marrow spaces between the bone trabeculae; (b) the resulting cement cuff exhibits an optimal wall thickness of about 2—3 mm; (c) the cement completely encompasses the metallic stem without any defects.

may also force the bone into a totally strange system of load transfer.

We classified the reaction of the bone to a cemented implant in three overlapping phases:

1. Firstly, we observed the repair of post-operative tissue damage and revascularisation of the bony implant bed.
2. The second phase was the adaptation of bone structures to the load transfer which lasted about 2 years.
3. Thirdly, an almost stable implant bed was formed which hopefully hosts the implant over the long run.

Microscopy of the adjacent bone removed soon after insertion of the cemented prosthesis revealed an infarction with necrosis of bone marrow. The dead marrow tissue was first replaced by fibrous tissue and then repair of the fractured trabeculae occurred within this fibrous tissue comparable to normal fracture healing. The necrotic bone trabeculae were removed by osteoblastic resorption and new bone was formed within the fibrous tissue.

Osteoid and subsequently mineralised bone penetrated the irregularities of the bone cement surface. Gaps were filled and protruding spheres of the bone cement were surrounded by osteoid or bone (Fig. 5.2a, opposite). The amount of direct bony contact to PMMA cement in serial tissue sections of a non-failed total hip replacement obtained at autopsy showed that 20% of the contact area in the femur consisted of bone with a little less in the acetabulum.

Years after implantation the bone which was formed during the remodelling process adjacent to the implant often does not calcify. The nature of the osteoid present at the contact areas, whether harmful or beneficial, is an area for discussion. Some workers assume that the presence of newly formed osteoid is caused by a disturbance of calcification induced either mechanically or chemically (e.g. by the monomer, catalysts or the X-ray contrast medium).

In areas of stable implant fixation where soft marrow tissue is bordering the bone cement, and sometimes also in areas where bone is close to but not in direct contact with the cement surface, foreign body giant cells can be seen together with a connective tissue membrane; intact haemopoietic bone marrow

---

Fig. 5.2a,b. The so-called permanent implant bed. Bone cement—bone border 4 years after implantation of a total hip replacement. **a** Two protrusions of a bone trabecula connect the surface of a cement cuff (*left side*, dissolved during PMMA embedding) with the spongy bone. While the trabecula itself is mineralised regularly (*green stain*) the contact areas remain as non-mineralised osteoid (*red stain*). Foreign body giant cells containing granulation tissue cover the major part of the cement surface. Undecalcified specimen, Pentachrome stain. **b** A broad seam of granulation tissue separates the bone marrow from the cement cuff (*left side*, dissolved during paraffin embedding). The structures of the cement surface are described by the contours of the foreign body giant cells and macrophages that filled the gaps between and covered the surfaces of the superficial PMMA beads. The granulation tissue consists mainly of broader strings of fibrocytes. PMMA pearls isolated from the cement cuff are found within this layer; they represent either tiny protrusions of the cuff or fragments separated due to mechanical stresses. Paraffin embedding, Azan stain

Fig. 5.3. Press-fit femoral stem fixation. Micro-radiograph of a cross section through the medial part of a Zweymuller cementless anchored femoral component made of TiA1V-forged alloy. The narrow planes of the rectangular cross section are orientated in a medio-lateral direction and are responsible for the load transfer. There the micro-radiograph demonstrates perfect implant to bone contact. The antero-posterior surfaces are in contact with bone marrow; circumferentially orientated bone trabeculae together with the cortical bone around the wedges may play a role in stabilisation of rotation.

Fig. 5.4. Bone ingrowth onto a titanium alloy surface. A Zweymuller femoral component is anchored by newly formed mineralised bone. This new bone bridged a gap between the old cortical bone of the implant bed (*outer right side*) prepared at operation and the implant's surface (*left side*). After having reached the implant, the bone structure spreads in order to achieve a better surface contact. The gap between bone and implant is thought to be artificial as no cells could be detected in this and many other areas of the preparation. Undecalcified preparation, Toluidine blue stain

Fig. 5.5a,b. Adaptation of the acetabular bone to a cementless anchored implant. Two stable fixed UHMW-polyethylene screw-type sockets in pre-cut acetabula. **a** 3 months and **b** 12 months after implantation. **a** Post-operative repair of tissue damage and remodelling of bone result in a loss of anchorage. Though the bony threads are mainly replaced by fibrous tissue, the tips of the thread as well as the top of the UHMWPE cup are still in contact with bone and secure the fixation. **b** New bone formation within the threads and closure of the trabecular structures by a new cortical layer on the top of the cup provide a stable implant bed.

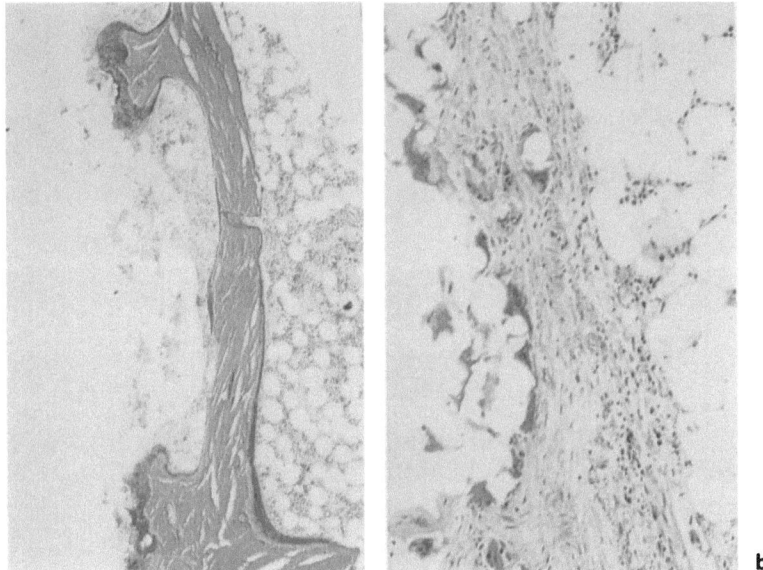

Fig. 5.2

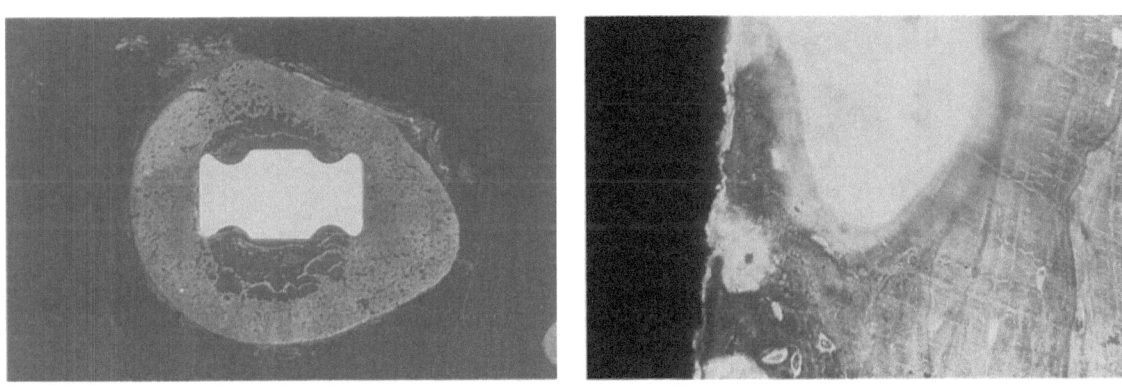

Fig. 5.3

Fig. 5.4

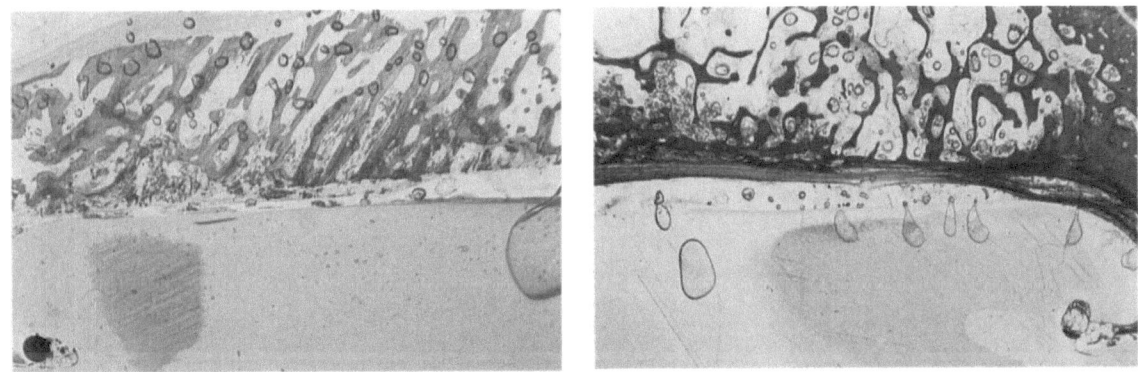

Fig. 5.5

is found beyond this membrane. This compound suggests a stable interface where no toxic or mechanical disturbances cause changes in morphology (Fig. 5.2b, opposite p. 28).

Flattened spheres of bone cement were observed in areas where the cement dough, cement pearls softened by the monomer, was firmly pressed against bone during insertion of the prosthesis. Together with the interdigitation into the spongy bone, this effect is an early indication of primarily firm anchorage of cemented implants.

However, the bone has to be remodelled in order to repair the implant bed and needs to adapt to the change of load transfer. Therefore the primary interlock is weakened and a delicate equilibrium between the amount of bone and applied stress has to be maintained. In the intertrochanteric area not only the number of bone trabeculae but also their predominating direction changes from rectangular to a more longitudinal direction. In specific areas, depending on the type of implant and shape of the cement cuff, we observe loss of mineralisation in cortical bone. These phenomena are thought to be principles of biomechanical adaptation which apply both to cemented and non-cemented implants. The implant-carrying bone is remodelled as well as every healthy bone. This is of significance for the interface between bone and the implant.

The regular remodelling of bone can be adversely influenced by infection, mechanical movement, instability, wear and corrosion products of the artificial bearing and by shattering and degradation of the bone cement cuff.

In unstable cemented joint replacements, we not only observe a delayed or defective mineralisation but also a more or less progressive replacement of bone and bone marrow by a space containing granulation tissue. The expansion of this membrane appears to occur at intervals. Semi-stable states in contact areas with micro-movements are characterised by scar formation within the fibrous tissues and the formation of synovial-like membranes. We think of the synovial-like membrane as a transformation of the reticular network of cells at the bone—cement interface which can easily be disturbed and become necrotic as soon as mechanical movement or pressure exceeds a specific degree.

The indications for the use of so-called press-fit prostheses implanted without bone cement are comparable to those for the use of cemented joint replacements.

The fundamental three stages of tissue reactions to implantation also apply. The healing process is not adversely affected by the curing bone cement and its leachable constituents; depending on the compatibility of the material used, a close contact with bone is possible.

One year after insertion of a cementless press-fit femoral stem made out of TiA1V-cast alloy, we found no periosteal reaction, very little loss of mineralised cortical bone and a direct contact between bone and the titanium surface (Fig. 5.3, opposite p. 28). The contact area increases with length of implantation. Microscopically, the gaps are filled with bone right to the titanium surface; nevertheless, some areas of marrow and fibrous tissue are left in contact with the alloy (Fig. 5.4, opposite p. 28).

Elasticity differences between bone and metal implants are responsible for relative movements. The type of prosthesis we have used for several years is primarily fixed in the diaphysis of the femur. Proximally, in the intertrochanteric region, we observe more fibrous tissue in contact with the implant which is interpreted as a result of movements not inducing pain but sufficient to prevent bone ingrowth.

The bone architecture around a non-cemented acetabular socket could be studied 3 months after implantation (Fig. 5.5a, opposite p. 28). The implant bed for this type of socket is a pre-cut thread where, depending on the contours of the acetabulum, minor areas of the subchondral cortical bone corresponding to the threads had been removed. The sample was processed with the implant in place and the histology revealed that the bone trabeculae were still orientated in a rectangular direction on to the former centre of the acetabulum. Another specimen taken 1 year after implantation showed the trabeculae to be completely covered by an internal lamella, and complete remodelling of the bone between the threads (Fig. 5.5b, opposite p. 28).

Yearly follow-up examinations of patients with non-cemented total hip replacements revealed excellent pain-free function and

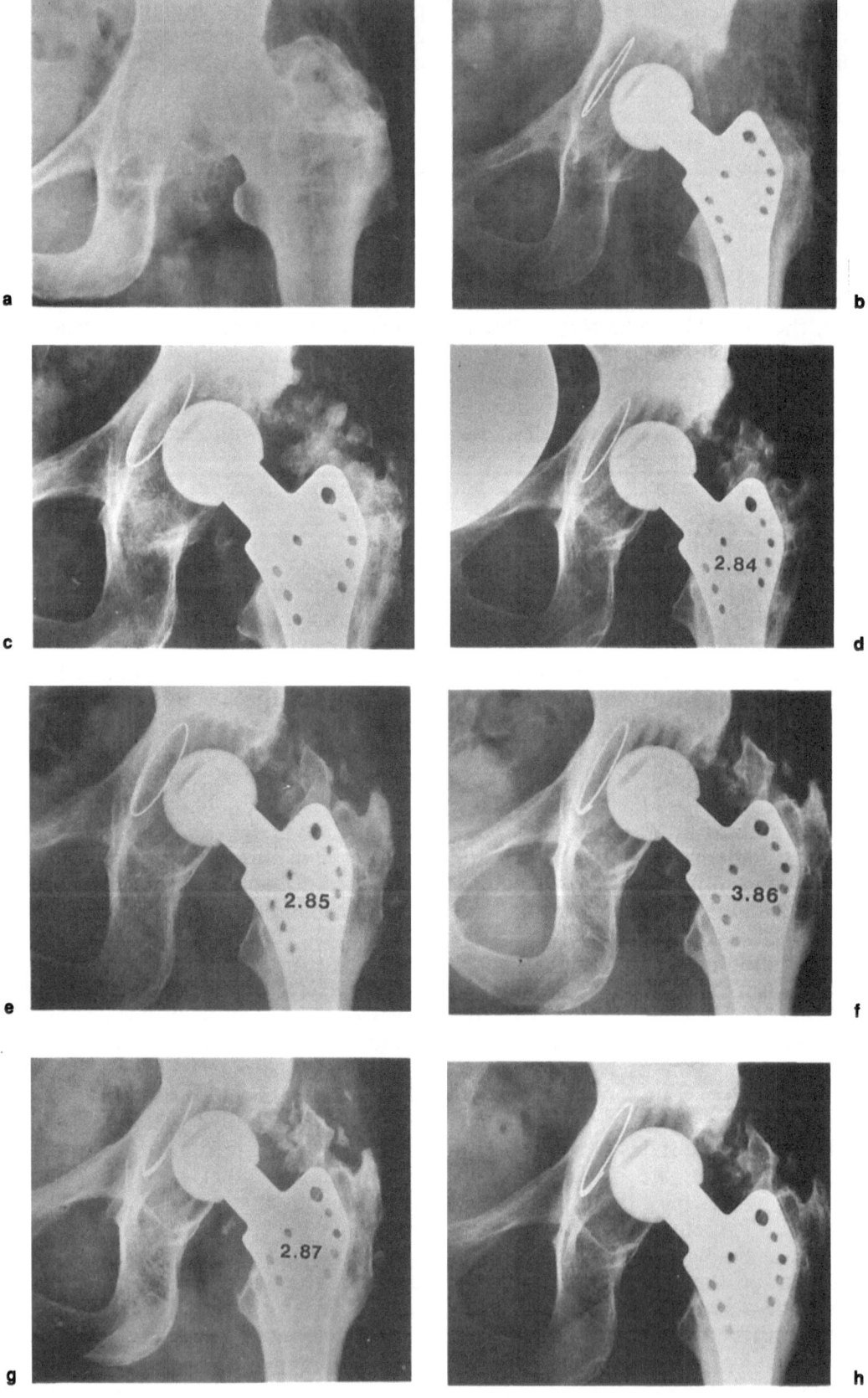

Fig. 5.6a–h. Adaptation of acetabular bone to a cementless acetabular replacement. Roentgenographic antero-posterior follow-up of a female patient with severe dysplasia of the left hip, previously treated with flexion osteotomy and implantation of a total hip replacement consisting of a screw-type all UHMWPW socket, cementless femoral stem (TiAlV-forged alloy) and Biolox ceramic ball head. **a** pre-operatively; **b** third post-operative day; **c** 6 months post-operatively; **d** 12 months post-operatively; **e** 24 months post-operatively; **f** 37 months post-operatively; **g** 48 months post-operatively; **h** 61 months post-operatively. The socket is screwed in a pre-cut implant bed; the threads are clearly visible 6 months after implantation. Remodelling of the former sub-chondral lamella results in loss of bone threads (**c**). Adaptation to the applied load by new formation of sclerotic bone in the threads after 12 months; moderate ectopic bone formation above the trochanter major region (**d**). Stable anchorage of the socket with good adaptation of bone for 5 years (**e–h**).

◄——

good adaptation of the bone to the implants, as shown in the radiographs (Fig. 5.6). Due to polyethylene wear against bone, these sockets have been replaced by pure titanium metal-backed UHMWPE sockets of identical design and these are showing the same satisfactory results. We expect these implants to function longer than the best cemented sockets.

Though the direct fixation of implants undoubtedly offers many advantages over cemented devices, for most of the patients we always have to remember that these implants too, depend on the delicate equilibrium of implant acceptance and tolerated forces. This implies that all afore-mentioned adverse factors, apart from those connected to the use of bone cement, may limit the time of stable implant fixation.

We should always bear in mind that the bone in which we implant our prostheses is not the same as the bone which has to carry those implants in the long term.

A cemented prosthesis has the best fixation right after implantation, but this fixation will only remain safe and survive the subsequent remodelling process and future adverse effects if the cement cuff is well distributed in the trabecular bone and diaphysis, and if optimum mechanical properties prevent fatigue fractures.

This note of caution has to be applied to the press-fit non-cemented prostheses too, but their advantage lies in the possibility that right from the outset they are anchored in such a stable way that bone remodelling cannot only occur afterwards without the danger of losing any of the implant, but can also lead to a stronger bond between implant and bone.

Acknowledgement. The author is indebted to Professor Dr F. Lintner, MD, University of Vienna, who prepared the tissue sections and micro-radiographs of the cementless joint replacements.

Discussion

The Chairman - **Mr Older**

The Panel - **Dr Draenert**
      - **Dr Linder**
      - **Dr Malcolm**
      - **Professor Willert**

Mr Older: Is there any significance in the giant cells?

Dr Malcolm: The presence of giant cells between the cement and living tissue was not a particularly striking feature. We saw them occasionally, more particularly between the cement and fatty haemopoietic marrow than between the cement and bone. I am not sure what significance I place on those giant cells except that they were present in small amounts, even in the specimens where there was no evidence of fibrous membrane 20 years after the implant.

Dr Draenert: There seems to be a very good correlation between the appearance of foreign body giant cells and micro-movements, according to our findings.

Dr Linder: In many of my specimens there was a direct contact between the bone and bone cement, and in some places there were, as you said, fatty marrow and giant cells. They occurred at the same time as we observed the direct bone cement contact.

Professor Willert: I cannot say very much about the importance and role of giant cells. They seem not to signify any major toxicity or disturbance. I observed that giant cells were very present around bone cement implants after 14 days. I think that the number of giant cells diminishes in the long run, with a predominance of fibrous tissue.

Professor Slooff: Years ago people said that there was never a direct contact between the cement and bone. Now we are told the reverse. Why are you talking about this direct contact? Have the techniques to establish direct contact improved or is the cement better than it was years ago?

Professor Willert: In 1972 we published our first paper on the reaction of direct contact. In our autopsy specimen, Denning found 20% direct bone contact, the rest soft tissue in a successful implant which shows us that such a contact is sufficient to bear the prosthesis.

Professor Slooff: Why, then, were you changing from a cemented prosthesis to an uncemented one?

Professor Willert: We changed in 1982 to the uncemented prosthesis because of the bone damage which we saw in the cemented prosthesis. To rectify it we did uncemented prostheses for 6 years. Now we use cement in people over 70 and uncemented prostheses in the younger patients.

Professor Slooff: Under 70?

Professor Willert: This is a little open-ended. I would say that we rely more on the biological than the chronological age.

Professor Slooff: I am not making a direct attack on you, I am just combining what all of you are saying and have shown. We were shown a three-dimensional reconstruction of the Charnley prosthesis. What you are saying is that, using the same prosthesis, you saw no signs of loosening in your specimens.

Professor Willert: I think we can combine the findings. Dr Draenert showed us the unsuccessful prostheses, I have seen only the good ones up to now! The clinical results of our investigations showed that this type of uncemented prosthesis is no worse than cemented. There is bone atrophy, some thigh pain and revisions are sometimes necessary due to wear of the polyethylene cup. But we are still using it, and the patients are still 85 to 90% satisfied with their prostheses.

I would take Dr Draenert's findings together with mine and say "This can happen with this prosthesis". The decision has to be based on the clinical results. Possible infection and all the clinical parameters have to be considered before the prosthesis can be judged a success or not.

Dr Draenert: It is very difficult because it cannot be discussed on a scientific basis. Comparing the results that we found and our three-dimensional reconstruction, we can say that the calculations completely fit the histological findings. The prosthesis clearly showed typical deformation of bone and press-fit type of anchorage. Certainly periosteal reactions were more or less pronounced. The findings of Linder are the same as ours, only the interpretation is different.

Dr Linder: I have a question for Dr Draenert. I have the impression that you said that in those cases where you have a very stable distal fixation of the prosthesis you saw fibrous membrane around the proximal part and at the same time a feeling of stress. Do you relate that to the development of the fibrous membrane?

Dr Draenert: I think that is a completely different problem. It might be handled better with histological results associated with osteosynthesis. Stress yielding will not mean loss of contact and the development of fibrous tissue.

Dr Linder: Is it just osteoporosis?

Dr Draenert: Yes.

Mr Northmore-Ball: Dr Albrektsson showed us a titanium screw going into bone, and said that because of the very small trauma in putting it in, it was always covered by fibrous tissue. He showed us a lovely series of pictures of this fibrous tissue gradually turning into bone and then becoming osseo-integrated. Are you saying that an osseo-integrated implant must always go through an initial phase of being covered in fibrous tissue?

Professor Albrektsson: Personally I do not think that there has to be a fibrous tissue interval, but that callus forms in mature tissue between the necrotic bone and the implant. I think also that at certain points there is direct contact between the necrotic original bone and the thread that holds the screw in place, during which phase that callus tissue matures into bone.

Professor Solomon: In those specimens where you saw macrophages and giant cells have you any histological evidence as to whether they have any osteoclastic activity or potential?

Dr Willert: We found giant cells and macrophages containing granuloma. These cells were loaded with foreign body particles which do not take part in bone resorption. There is a distinct gap with loss of fibrous tissue and vessels. Other cells then are responsible for bone resorption and formation of new bone.

Professor Solomon: Are bone resorption cells present? If there is a substantial fibrous membrane, presumably bone must have been resorbed. What is causing the bone resorption?

Professor Willert: The presence of a fibrous tissue membrane does not mean that the bone will be resorbed histologically. Bone resorption is only brought about by osteoclasts. Normally we only see osteoclasts at the interface during remodelling. Osteoclasts appear if there is some reaction, such as infection, bleeding and storage of foreign body particles within this fibrous membrane. I only interpret what I see as bone resorption when there are osteoclasts. This could clearly be seen around a granuloma developing at the bone cement or bone interface, for instance. This area has to be investigated histochemically.

Professor Slooff: If the fibrous membrane is not caused by bone resorption, then what is the cause?

Professor Willert: Fibrous membrane does not cause bone resorption by its existence. I can only speculate about the formation of the fibrous membrane. These are the areas where bone is not in contact with the surface of the cement or the spaces in between. There is

always a cover, mantle or sheath between the bone marrow and the implant.

Mr MacDonald: Are bone trabeculae commonly found directly against the polyethylene surface of the acetabular socket?

Mr Older: We have heard a lot from you all about titanium on the metal side of the interface. Is it just by chance that your research involves titanium, or do you have any experience of stainless steel components? If you do, do you see any difference between titanium and stainless steel? From your experience of the microscopy of the interface, what do you see as the significance of the conversion film on the implant?

Dr Draenert: I can contribute only one specimen, an operation performed in Basle, where they tried to test titanium screws in comparison with steel. Every second screw was titanium. Ten years after the operation the results were identical. Both types of screw showed the same excellent contact between metal and bone.

Mr Older: What has been your experience, Dr Linder?

Dr Linder: A study has shown fibrous membrane between screws made of stainless steel and titanium alloy. An electron microscopic study did not detect any difference in the cellular reaction to the two metals in direct bone—metal contact but no difference could be seen in the soft tissue.

I think that Dr Albrektsson knows more about this than I do. Charnley's book showed evidence of a direct contact between a stainless steel screw and bone. I think there is no doubt that the screw had a direct bone implant contact. But I think Charnley was of the opinion that these screws were anchored in such a way that they did not really carry any load. If they were to carry load they would subsequently loosen. That might have been the prevailing opinion in those days.

Professor Willert: I can only quote the work of Steinmann, who showed superior behaviour of the surface biochemistry and bio-electricity of titanium at the ion level. He thinks that this is the reason why titanium and titanium alloys seem to be more attractive for bone ingrowth than cobalt and its alloys.

Mr Smith: We are investigating the interface of screws in the knees, the femur and acetabulum around the hip and seem to be drawing the same inference from each of the sites. Does the panel think that there is any difference between these screw sites? Are you trying to extrapolate from what you are seeing instead of concentrating on one site at a time?

Professor Willert: There are at least two problems involved. One is mechanical and the other is biological. The problems on the level of cell interaction are the same in all the sites, but the biomechanical problem is different. I do not know to what extent these differences influence the tissue reaction.

Dr Linder: I think also there is no reason to believe that a bone cell in the acetabulum is biologically different from a bone cell in the tibia. But as the biomechanical situation differs, so does the influence on the cell during the healing phase of the implantation. What you are saying is probably true. We should treat these different parts differently in order to achieve optimal results.

Mr Gruebel-Lee: Between 1964 and 1966 we used titanium with Austin—Moore prostheses. Many of them did well, and were stable at post mortem. If there was movement and resorption of bone, the prosthesis had to be removed and we always found profound dark grey staining of the tissues. Titanium is alright if it works but can be disastrous if it fails.

Chapter 6

Tissue Reactions to Metallic Biomaterials

T. Albrektsson

Commercially pure titanium has some advantages in cranio-facial reconstructions when compared to other metals. Use of commercially pure titanium oral implants has been well documented over a 20-year period (Albrektsson and Lekholm 1989). Even with skin-penetrating implants there is now a positive clinical follow-up of more than 10 years (Albrektsson et al. 1987).

There is considerable load bearing with dental implants. A man has only two hips but he needs a great many dental implants which, scientifically, is a great advantage. The total experience worldwide of titanium screws inserted in the upper and lower jaws is now 250 000 implants, and the longest follow-up is 24 years, so there is some clinical experience, albeit not from the orthopaedic field, to back up my otherwise basically science-oriented statements.

Any material to be used in the human body needs careful pre-evaluation from a technical and biological viewpoint. Animal experiments must be conducted. The ultimate testing situation must always be clinical, however, and it is only possible to transform clinical experience from the field of cranio-facial reconstructive surgery in part to that of orthopaedic surgery. Here our current knowledge is limited to dynamically loaded canine knee joints that remained functional for 2 years with direct bone-to-implant anchorage and no ten-

dencies to migration of the anchoring devices (Albrektsson et al. 1989; Carlsson 1989; Rostlund 1990).

Tissue Reactions

I strongly believe that when we talk about tissue reactions to various biomaterials we have to base our assessments not only on qualitative but also on quantitative evaluations. We look at retrieved clinical implants in our laboratory and are able to calculate the percentage of direct bone-to-implant contact (Fig. 6.1a,b). What is the scale, therefore, when we talk about major or minor tissue reactions? Radiography, the most sensitive tool of the clinician, has a poor resolution level. The radiologist, if very skilful, may be able to describe accurately to a size of 100 μm but it is totally impossible to see radiographically if an implant is directly bone anchored (Fig. 6.2a,b) because the soft tissue cells are only 10 μm in size. Histologically, there are evident differences between various metallic biomaterials if the resolution level is increased to the light and electron microscopical levels. It may be that the clinical reactions we see after 10 years because of corrosion could, therefore, be seen by the biologist after a few

a

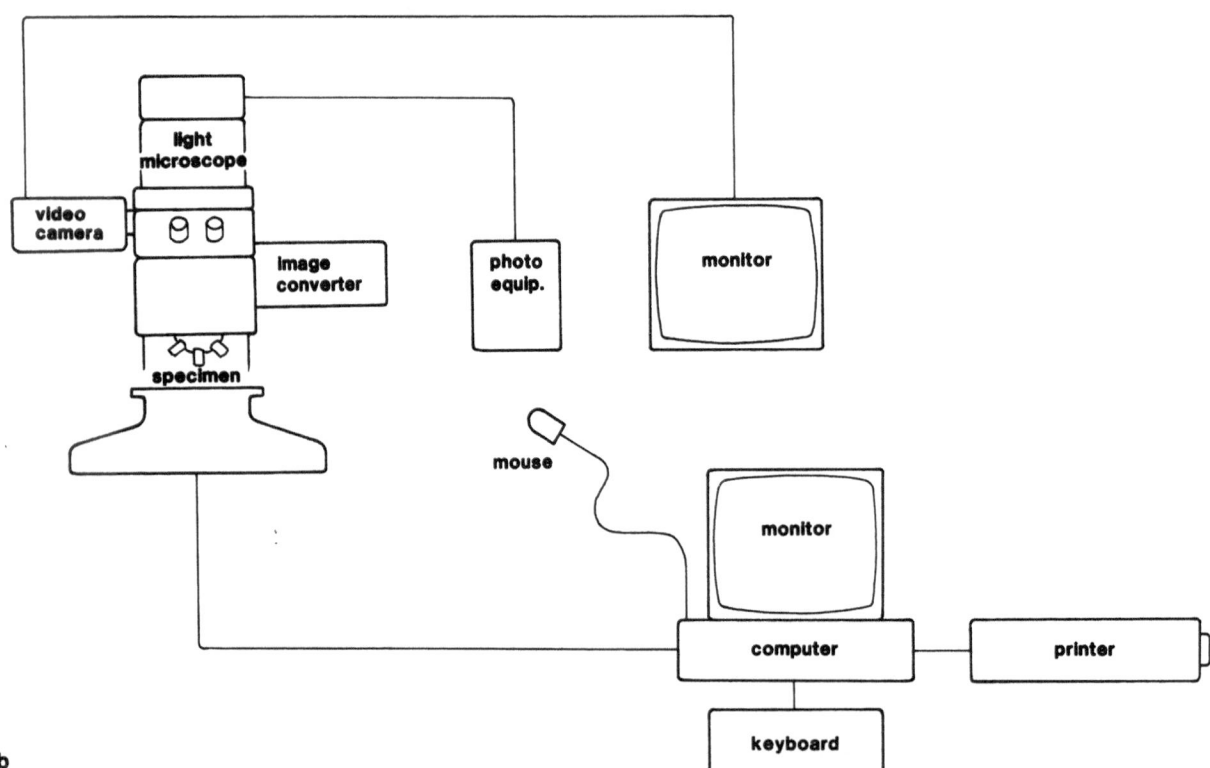

b

Fig. 6.1. Leitz Microvide equipment for calculations of bone to implant contact zones (**a**). Plan of the set-up (**b**).

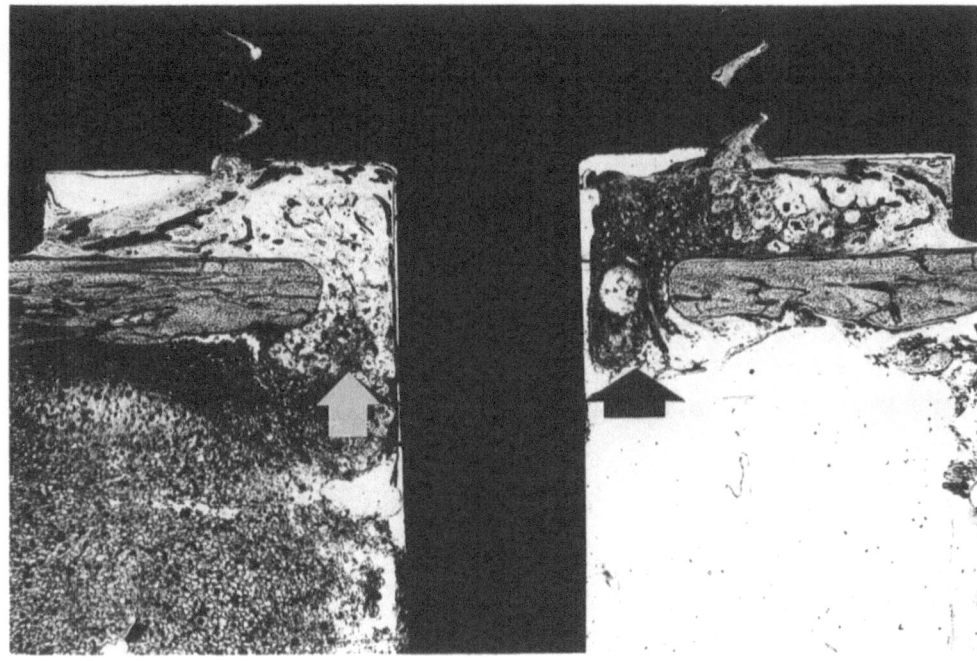

Fig. 6.2. a Histology of titanium implant inserted in bone revealing a clear zone of fibrous tissue (*arrows*). **b** Radiography of the same implant demonstrating a false image of a direct bone-to-implant contact.

months if more sensitive methods than radiography were available.

Using electron microscopic techniques, we looked at the interface between bone and various metals (Fig. 6.3). In the specimens containing pure gold, no bone contact was seen (Albrektsson et al. 1982). There were varying responses to stainless steel, although the proportion of bony contact has been less than that reported with commercially pure titanium (Albrektsson and Hansson 1986; Linder et al. 1989). In the case of experimental implants of commercially pure titanium, there were hydroxyapatite crystals in the interface at the resolution level of the electron microscope (Hansson et al. 1983). Titanium, niobium and tantalum gave rise to very similar reactions when they were com-

pared (Johansson et al. 1989b), but we do not know for certain whether these favourable tissue reactions to the cited metals are exactly replicable with clinical implants of bulk metal of another elastic modulus (Albrektsson and Sennerby 1990).

Recent findings, however, indicate that there are significant differences between titanium and titanium—aluminium—vanadium alloy (Johansson et al. 1989a,c; 1990). When we tried to remove implants after 3 months, those made of commercially pure titanium were always more difficult. Direct contact in places and soft tissue in other areas could be seen with commercially pure titanium when we checked the amount of bone in the interface. This general finding was similar with the titanium—aluminium—vanadium alloy.

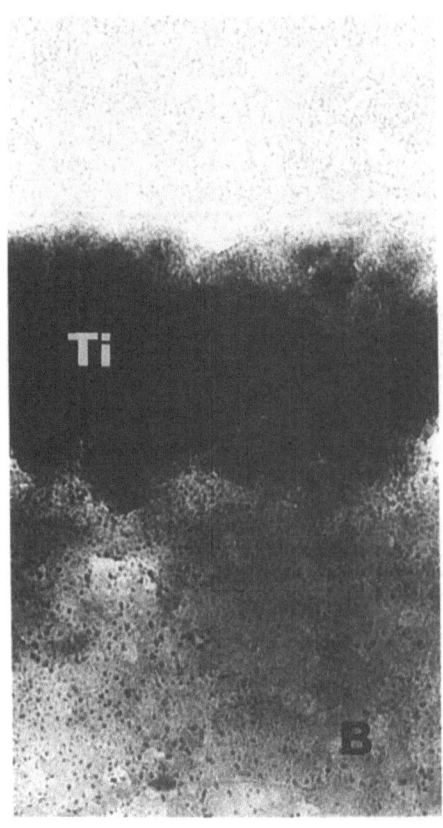

Fig. 6.3. Titanium (*Ti*) to bone (*B*) interface as seen in electron microscopical investigations of plastic plug implant.

When measured quantitatively, however, there was a significant difference in the amount of bone seen at the interface in favour of commercially pure titanium (Fig. 6.4a,b). We are not certain of the reason for this tissue difference but proven leakage of vanadium and aluminium from the specimens is suspected. Leaked-out aluminium may compete with calcium during the calcification process, resulting in a local type of "osteomalacia" (Johansson et al. 1989c).

Experiments have indicated that a true integration may exist between titanium and bone tissue in the form of a chemical interaction between the two (Steinemann et al. 1986). Steinemann's definition of osseo-integration is biomechanical — resistance to shear forces as well as to tensile forces, which he was able to check separately. Implants were less stable some time after implantation due to interfacial bone resorption, and then the shear strength increased rapidly up to 25 newtons/mm². These findings were not unex-

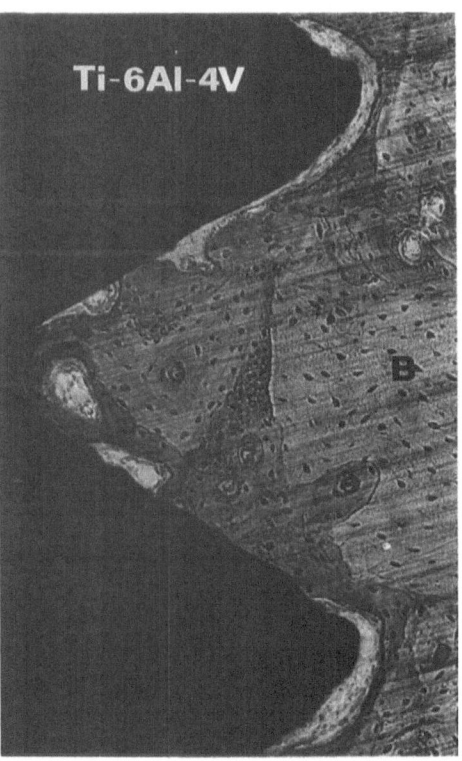

Fig. 6.4. Commercially pure titanium (*Ti*) showing a much greater bone-(*B*)-to-implant contact (**a**) than a Ti-6Al-4V alloy implant (**b**).

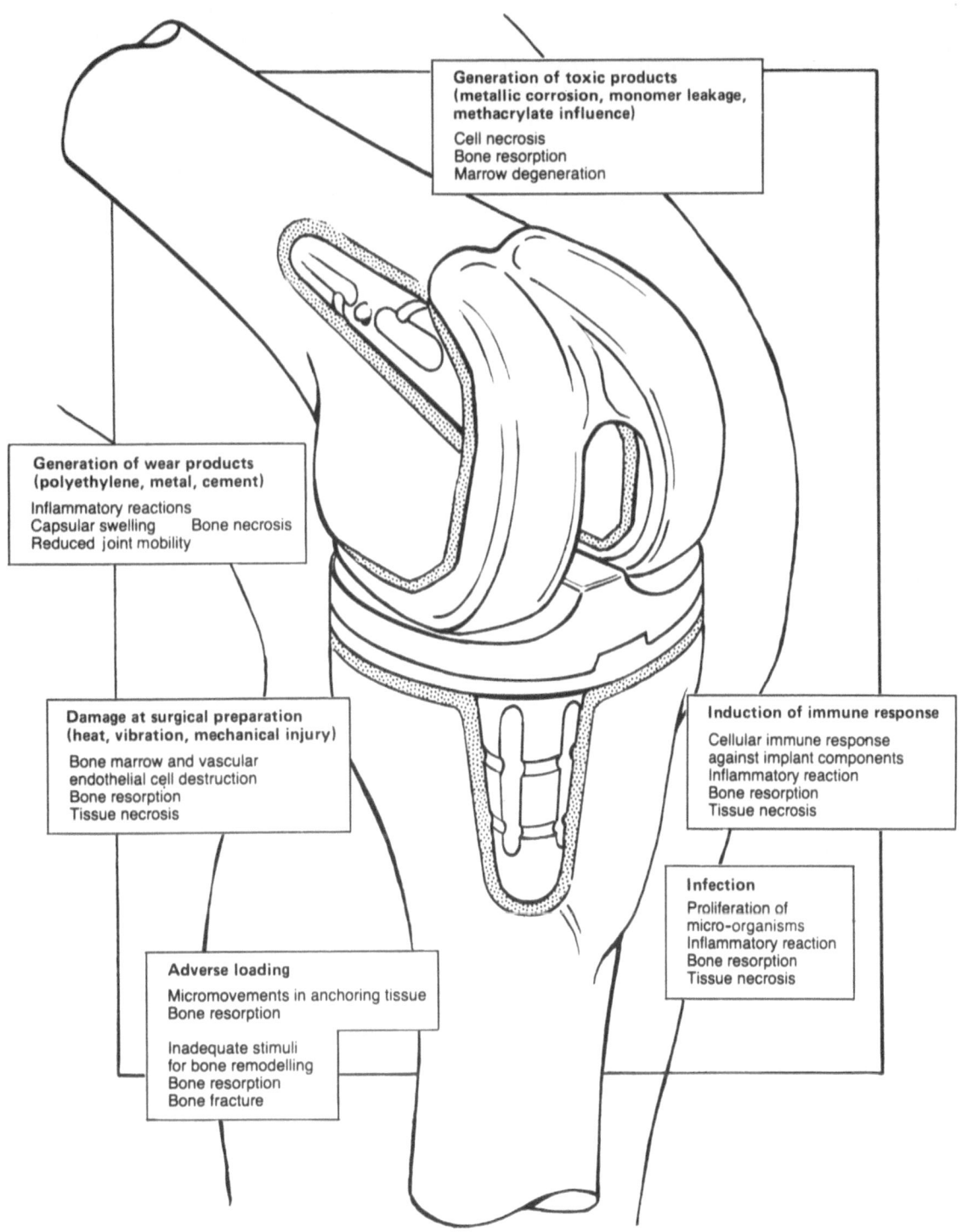

Fig. 6.5 Biological reasons for failure of conventional type of knee replacement.

pected, but the observation of a resistance to tensile forces was of specific interest. If there is a ·true controlled experimental situation and pure tensile forces can be measured separately, this can only be explained by some type of true interaction between commercially pure titanium and bone tissue. The experimental work showed no resistance to tensile forces during the first 100 days after implantation, but thereafter clear resistance was observed. This has been explained as a chemical interaction between the titanium oxide that inevitably covers the titanium implant and the bone tissue (Steinemann et al. 1986). To my knowledge, this is the only paper demonstrating such a contact.

Are there any negative side effects of a metal such as commercially pure titanium? A blackening reaction is sometimes seen around clinically used titanium implants. We strongly believe that such local leakage of titanium is dependent, at least in part, on how the implant is treated prior to insertion (Albrektsson 1985). The manufacturing and control of the surface conditions during sterilisation of the implant are relevant but these may not be the only crucial parameters.

Titanium—aluminium—vanadium alloy and commercially pure titanium implants were inserted into baboons for up to 8 years. Vanadium was not found to leak out to any significant level. Evidence was found, however, that the titanium levels became elevated in the lung and spleen after insertion of a titanium implant, but there was no evidence of biochemical or haematological toxicity because of it (Woodman et al. 1984). Nevertheless, more and more titanium is found in the lung tissue of the baboon up to roughly 3 years after implantation, and then there is a plateau phase. I have looked in the literature for adverse tissue reactions due to titanium, but was only able to find one report in a dermatological journal of an assumed allergy to the metal (Peters et al. 1988).

A continuous increase in leakage of aluminium in the case of titanium alloys was reported (Woodman et al. 1984). We know that if it continues this will be problematic sooner or later. Aluminium is neuro-toxic and there are problems if it reaches neural tissue, but we do not know when hazardous levels are found in the case of leakage from an aluminium-containing implant. If this leakage occurs after 100 years or more, the question is entirely academic, but I would still hesitate at present to use aluminium-containing titanium because of the negative local bone reactions to titanium alloys.

I also remain hesitant about direct bone anchorage with existing hip and knee prostheses since it is very difficult to control simultaneously the several parameters necessary for achieving so-called osseo-integration (Albrektsson et al. 1981). The surgical technique is one of those parameters which is very difficult to control with traditional types of orthopaedic implants so that the interfacial bone temperature remains in the range of 47°C or lower during the surgical intervention. The technique to achieve proper bone anchorage of an implant is so sensitive that poor control of the surgical trauma is sufficient for the implant to be permanently anchored in soft tissue, irrespective of the most well-controlled metallic biomaterials being used. There are a variety of biologically based reasons for failure of hip or knee joint implants as summarised in Fig. 6.5.

Acknowledgement. Professor Albrektsson, who presented this paper, is indebted to his colleagues for their valuable support in their combined research efforts.

References and Further Reading

Albrektsson T (1985) The response of bone to titanium implants. CRC Crit Rev Biocomp 1:53—84

Albrektsson T, Hansson H-A (1986) An ultrastructural description of the interface between bone and sputtered titanium or stainless steel surfaces. Biomaterials 7:201—205

Albrektsson T, Lekholm U (1989) Osseo-integration — current state of the art. Dent Clin North Am 33:537—554

Albrektsson T, Sennerby L (1990) Direct bone anchorage of oral implants. Clinical and experimental considerations on the concept of osseo-integration. Int J Prosth 3:30—41

Albrektsson T, Branemark P-I, Hansson H-A, Lindstrom J (1981) Osseo-integrated titanium implants. Requirements for ensuring a long-lasting, direct bone anchorage in man. Acta Orthop Scand 52:155—170

Albrektsson T, Branemark P-I, Hansson H-A, Ivarsson B, Jonsson U (1982) Ultrastructural analysis of the interface zone of titanium and gold implants. Adv Biomat 4:167—177

Albrektsson T, Branemark P-I, Jacobsson M, Tjellstrom A (1987) Present clinical applications of osseo-integrated percutaneous implants. Plast Reconstr Surg 79:721—730

Albrektsson B, Albrektsson T, Carlsson L, Rostlund T (1989) The bone-anchored knee replacement. In: Albrektsson T, Zarb G (eds) The Branemark osseo-integrated implant. Quintessence, Chicago Berlin Tokyo Sao Paulo, pp 251—255

Carlsson LV (1989) On the development of a new concept for orthopaedic implant fixation. Thesis, University of Gothenburg, Gothenburg, Sweden. Elsevier, Amsterdam

Hansson H-A, Albrektsson T, Branemark P-I (1983) Structural aspects on the interface between tissue and titanium implants. J Prosthet Dent 50:108—113

Johansson CB, Albrektsson T, Thomsen P (1989a) Removal torques of screw-shaped commercially pure titanium and Ti-6A1-4V implants in rabbit bone. Adv Biomat (in press)

Johansson C, Hansson H-A, Albrektsson T (1989b) A qualitative interfacial study between bone and tantalum niobum or commercially pure titanium. Biomaterials (in press)

Johansson C, Lausmaaa J, Ask M, Hansson H-A, Albrektsson T (1989c) Ultrastructural differences of the interface zone between bone and Ti-6Al-4V or commercially pure titanium. J Biomed Eng 11:3—8

Linder L, Obrant K, Boivin G (1989) Osseo-integration of metallic implants. II. Transmission electron microscopy in the rabbit. Acta Orthop Scand 60:135—142

Peters MS, Schroeter AL, van Hale HM, Broadbent JC (1988) Pacemaker contact sensitivity. Contact Dermatitis 11:214—219

Rostlund TV (1990) On the development of a new arthroplasty. With special emphasis on the gliding elements in the knee. Thesis, University of Gothenburg, Gothenburg, Sweden

Steinemann SG, Eulenberger J, Maeusli PA, Schroeder A (1986) Adhesion of bone to titanium. In: Christel P, Meunier A, Lee AJC (eds) Biological and biomechanical performance of biomaterials. Elsevier, Amsterdam, pp 409—418

Woodman JL, Jacobs JJ, Galante JO, Urban RM (1984) Metal-ion release from titanium based prosthetic segmental replacements of long bones in baboons: a long term study. J Orthop Res 1:421—432

Chapter 7

Osseo-integration of Metallic Implants in Animals and Humans

L. Linder

This is a review of some experiments that have been done on osseo-integration in animals and human beings.

A direct bone—implant contact is not a very new finding. Several people in the 1950s and 1960s reported on occasional cases of direct implant—bone contact and this was confirmed by Charnley (Charnley 1979). A stainless steel screw was found in contact with the bone. Most people thought at the time that bone could come in contact with an implant but could not bear load for long. Branemark has shown very clearly that integrated screws of titanium can, in fact, carry physiological loads over very long periods of time without loosening, and that is a major contribution (Branemark et al. 1977).

Why is it that the uncemented metallic implants in clinical practice today are not integrated in the majority of cases? In fact, it seems that in almost all retrieval studies they seem to be bordered by connective tissue. Is there a difference due to the implant materials? Pure titanium, which is used by Branemark in dental surgery, is not yet used widely in joint replacement. Instead, the implant materials in general use today are the chrome—cobalt alloys and titanium alloys. Stainless steel may not be used in uncemented joint replacement but it is a very common orthopaedic material.

Animal Experiments

It is possible to compare the bone reaction to implants of different materials in animals and to standardise a situation where the surgical trauma and the loading conditions are equal for all implants. Only the implant material is changed.

We compared implants of pure titanium, tivanium, vitallium and stainless steel (Linder 1989). The surface finishes were clinically used surface finishes so they were not identical. On the X-rays, there was a slight periosteal bony proliferation after implantation and an endosteal bony proliferation around the narrow part of the implant (Fig. 7.1).

Histological Techniques

The implant was removed together with the bone and embedded in plastic. Sections of this plastic were removed from the implant surface and after some practice it was possible to obtain clean fracture surfaces. The plastic now contained all the interface tissue and it was easy to section on any microtome.

Results

We studied 78 implants in 39 rabbits. Only three did not become integrated; two were of pure titanium and one was stainless steel, so all of the tivanium and vitallium implants became integrated. I see the results as a biological principle, that a tolerable implant material inserted under tolerable conditions will become osseo-integrated.

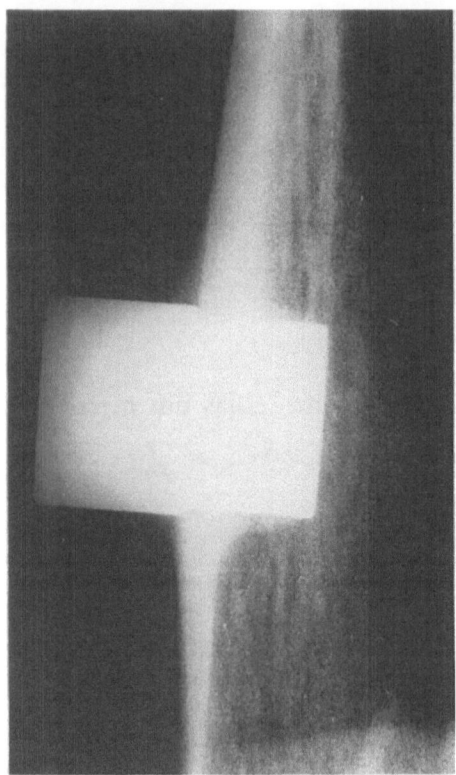

Fig. 7.1. An osseo-integrated implant in the rabbit tibia. There is periosteal and endosteal bone proliferation with no radiolucent line between implant and bone.

Choice of Metals

Using electron microscopy, osteocytes could be seen lying very close to the implant surface (Linder et al. 1989). There was no soft tissue or cellular layer but a direct contact between bone and implant. All four implant materials gave a similar picture. It would seem that the choice of metal is not the crucial factor in osseo-integration, even at this level of resolution.

Perhaps the choice of material is not the main reason why clinical implants do not become osseo-integrated. Bony ingrowth into the pores of the implant and bony contact with the implant surface has been reported in the literature for both titanium alloy and chrome—cobalt alloy. So the conclusions from the animal experiments are supported by the findings in some clinical biopsies. If the nature of the implant material is not critical, is it the local condition of the adjacent tissue that affects the healing process? The arthrosic disease itself may influence this tissue and be crucial to healing and fixation.

Methods

In a clinical experimental study (Linder et al. 1988), we inserted screws of pure titanium, using Branemark's method, into volunteer patients who were on the waiting list for a knee replacement (Fig. 7.2). Implants were inserted into 14 patients with osteoarthrosis and 11 patients with rheumatoid arthritis.

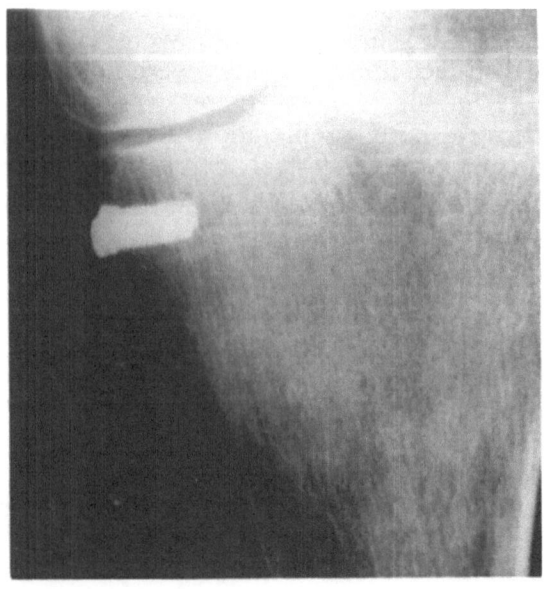

Fig. 7.2. The placement of the titanium screws in the human experiment is shown.

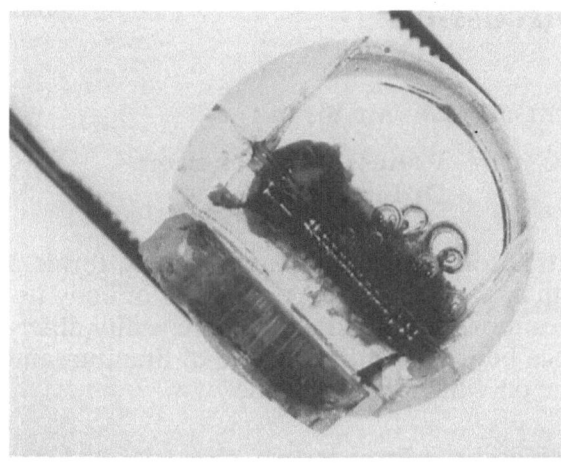

Fig. 7.3. A clean separation can be obtained by embedding implant and surrounding tissue in plastic and by subsequently separating the plastic from the implant. In such a case the intact interface tissue is preserved in the plastic and can be sectioned, even for electron microscopy.

The observation times varied between 1 month and 2 years owing to the nature of a waiting list. At the time of knee replacement the screws were removed either by a drill to obtain specimens which we embedded and separated from the implant (Fig. 7.3), or the entire resected tibial plateau was embedded so that we could cut through the intact metal—bone interface (Fig. 7.4).

We compared the histology in two patients, one with osteoarthrosis, the other with rheumatoid arthritis, both with 1 month observation time. There was excellent bone quality in the osteoarthritic patient with only

an island of necrotic bone still in contact with the tip of the screw thread. There seemed to be a remodelling of the whole screw thread with osteoid as well as viable bone in its depth. The rheumatoid patient, however, had one screw thread completely empty and another contained mostly soft tissue with only a little bone. There was quite a difference between the bone quality of these two patients.

There were 21 further implants, all with observation times of 5 months or longer. There was osseo-integration in all except one, in which the bone was not in contact with the screw but separated from it by a thin layer of fibrous tissue. There was also some fibrinous amorphous material in the depth of the screw threads and many foreign body giant cells.

In all the other cases there was direct contact between mineralised bone and screw. The interface was variable however, with direct contact between the bone and the implant in some places, and marrow tissue with occasional foreign body giant cells at other sites. We do not know yet what these giant cells are doing.

The general architecture of the trabeculae was not influenced at all, so that in cases of osteopoenia there were very few contacts between bone and screw. We observed condensation of bone along the interface in only very few cases, regardless of observation time.

Summary

Radiologically, titanium seems to behave as a neutral material. I think the results show that the biological requirements for osseo-integration are clinically present in orthopaedic surgery, so why aren't today's implants osseo-integrated? That is a very difficult question.

Today, uncemented implants are in a developmental period. Dental bridges in the upper jaw have proven successful with increasing experience (Branemark et al. 1977). Implants in the lower jaw have always been extremely successful, but even an osseo-integrated implant can loosen. Atraumatic surgery must be followed by atraumatic prosthetics. If osseo-integration is achieved, it must be

Fig. 7.4. The embedded and divided medial tibial plateau with a titanium implant in place is shown.

maintained, and it cannot be maintained unless the implant or the prosthesis is properly designed. Whether it is cartilage, fibrous tissue or bone, the biological reaction must be in balance with the biomechanics.

Conclusions

I conclude from these investigations that osseo-integration as a principle is possible in human beings. Osteopenia (or osteoporosis) does not preclude osseo-integration, but is a risk factor both during the healing period if loads are applied, or after the healing period, even if osseo-integration is established. Titanium is certainly an inert material. I have not been able to see any stimulation to bone formation, but this is extremely difficult to determine in absolute terms.

From the more philosophical point of view, my conclusion would also be that osseo-integration is an expression of the healing capacity of bone. We must learn more about the signals which govern tissue differentiation. The use of pure titanium is, however, no guarantee of osseo-integration. An osseo-integrated titanium implant can very well loosen, and there have been times when I have not been able to achieve osseo-integration even if I used pure titanium under very controlled and optimal conditions. Cement-free fixation in general is not synonymous with osseo-integration, and I feel that these terms should be separated at the present stage of knowledge.

References and Further Reading

Branemark P-I, Hansson BO, Adell R, Breine U, Lindstrom J, Hallen O, Ohman A (1977) Osseo-integrated implants in the treatment of the edentulous jaw. Scand J Plast Reconstr Surg 16:1—132

Charnley J (1979) Low friction arthroplasty of the hip. Springer-Verlag, Berlin Heidelberg New York

Linder L (1989) Osseo-integration of metallic implants, I. Acta Orthop Scand 60:129—134

Linder L, Carlsson A, Marsal L, Bjursten LM, Branemark P-I (1988) Clinical aspects of osseo-integration in joint replacement. J Bone Joint Surg (Br) 70:550—555

Linder L, Obrant K, Boivin G (1989) Osseo-integration of metallic implants, II. Acta Orthop Scand 60:135—139

Discussion

The Chairman - **Mr Elson**

The Panel - **Professor Albrektsson**
         - **Dr Linder**

Mr Elson: If titanium immediately forms a film as soon as it is exposed to air or body tissues, how do you know if there is any difference between a pure surface of titanium and one on which there is oxide?

Professor Albrektsson: There is evidence that titanium ions cause problems. It is a very reactive metal. I am still sure that the biocompatibility of the commercially pure titanium is very dependent on the oxide layer.

Mr Elson: Why does the plateau of titanium concentration in the lungs tail off? Presumably it is constantly being produced at the surface.

Professor Albrektsson: I do not know. Woodman reported a plateau in the case of titanium leakage into the lungs, but I cannot speculate on the reason.

Mr Elson: But the aluminium goes on for ever?

Professor Albrektsson: For 8 years in Woodman's study.

I have a question for Dr Linder. Do you think there is any difference between the behaviour of the lower jaw bone compared with other bones in the body? I am concerned about infection. It is remarkable what one can do to the membrane bones of the jaw. The dentists are surprised at what they are able to do, often operating under strange circumstances. Osteomyelitis of the lower jaw is unusual after dental procedures. Do you think this also reflects on the way in which screws behave differently compared with those that are put into the knee?

Dr Linder: I am not sure, but it is difficult to believe that the reaction of bone cells in one region would be dramatically different from another region. Perhaps the reason why these implants are not infected reflects the fact that they are integrated, and that osseo-integration itself acts as a biological seal for bacteria.

Mr Elson: A barrier to bacteria?

Dr Linder: Yes. Looking at the structure of the tissue around osseo-integrated screws in the jaws, there is some degree of inflammation because the gingiva is communicating with the oral cavity. There is a collagenous cuff around the titanium implant which contains the inflammation and there is no reaction at all in the bone. The implants are inserted into the jaws in a very different manner from joint prostheses. They are usually inserted in two stages and the tissues allowed to heal before any disturbing factor is superimposed, whereas hip and knee implants are inserted as one procedure.

Mr Elson: So there is no fundamental tissue difference between the two?

Professor Albrektsson: I do not agree that there is no difference. One type of load pattern differs from another. In one bone site we have one type of blood flow. We know that the proportion of blood is not evenly distributed in the long bones of the body. On the contrary, there is a great difference. Bone integration can clearly be achieved clinically and experimentally in the lower jaw. We can use a relatively easy one-stage procedure in the lower jaw with a good success rate; not so in the upper jaw. Among our canine cases we find a very similar reaction in the femur and the tibia. In the distal femur we can use a one-stage procedure mostly with good clinical results. We almost always fail to achieve the bony anchorage that we desire when we use a one-stage surgical procedure in the tibia. So we do see a difference, although I agree that the osteocytes are the same.

Mr Atkins: I think the fact that aluminium alloy seems to inhibit bone formation could be associated with the aluminium ions being toxic to osteoblasts as, for example, in renal osteodystrophy and dialysed aluminium-containing water. Do you think it may be a simple effect of aluminium ions being directly toxic?

Professor Albrektsson: Yes, we speculate on that. At present we are trying to measure whether aluminium has leaked out in the tissues with ion probe analysis. I think it is likely to be the toxic effect of the aluminium, but I do not know.

Mr Ling: Is it possible that aluminium actually competes for calcium in the bone, so that you have an osteomalacic situation which cannot be good mechanically?

There was a recent paper suggesting that vanadium inhibits apatite formation. Work on titanium alloy implants and the effects of the ions coming out of those implants when they are mobile was presented by the Hospital for Special Surgery at the combined meeting of the British and Greek Orthopaedic Associations in May 1989. There was a similar study from Heatherwood, England, of a series of titanium alloy femoral components that had started to loosen and then liberated large amounts of metallic material, so presumably there could be several local effects. The implants in both series were very small and did not span large segments of bone, whereas in the femur, at least with the intramedullary type of fixation, we are using implants that span a very large section of bone. That bone is really moving round the implant. So this is another dimension which is quite different from the osseo-integration of a screw in the jaw or a single screw in the upper tibia. The bone itself is moving around the device, so you might expect different histology in different parts of that implant. What do you feel about that as another aspect of this problem?

Professor Albrektsson: In reply to the first part of the question concerning the aluminium and vanadium leakage problem, I agree that there could be competition between aluminium and calcium.

With regard to vanadium leakage, investigators in the past have come to the conclusion that although vanadium is very toxic, its leakage is so small that it is taken care of, as far as we know, by the body's defence mechanism.

Dr Linder: With regard to the large implants, in a porous-coated hip prosthesis that was revised because of thigh pain, there was one kind of histology at one end and another at the other end. Histology showed there was some degree of ingrowth in the proximal part but there was no bone ingrowth lower down. There was just fibrous tissue round the smooth part of the PCA stem.

Professor Albrektsson: I totally agree. The reason why some of us want to design bone integration is we believe that if there is a great deal of bone present and it is well maintained, it would continue to remodel according to the loads put upon the device. If so, it would be advantageous to a soft tissue interface. I believe that it is very difficult to achieve and maintain osseo-integration with the present stem designs. That is why we do not intend to work with stem designs in our clinical implants.

Mr Gruebel-Lee: If you have a fracture into a tooth socket and the tooth is loose in the mandible, rapid onset osteomyelitis or sometimes a big fibrous membrane and non-union may result. I suspect that the success of a mandibular prosthesis depends upon its original fixation and rigidity.

Mr Macdonald: What are the stress levels that operate at the interface with dental implants?

Professor Albrektsson: There are considerable loads, but I am no dentist so that is not a very scientific response to your question.

Dr Grobbelaar: We have an ongoing test in our department, but instead of rabbits we used the South African baboon, a primate. In these cases there is a difference in tolerance to the three metals. We have no doubt that titanium alloy is a much more inert material.

There is a great deal of controversy at the moment about aluminium. In theory, aluminium actually advances the inertness of the material in the sense that it will bind to the hydroxyapatite and make the material more insoluble, so that the osteoclasts find it more difficult to dissolve bone.

We now do cancellous bone transplants in our unit routinely. When we implant cement we just screw in acetabulum cups. We have the impression that much better integration is achieved both clinically and radiologically. Would you comment on this?

Dr Linder: Did you say that you had better integration of the titanium alloy?

Dr Grobbelaar: We only use 6% aluminium/ 4% vanadium in all our cases; we do not use the pure titanium.

Dr Linder: I did not find any difference quantitatively between these materials. My method of analysis is far less sophisticated than that of Dr Albrektsson. I use cylindrical implants and he is using screws which give a large surface area, and perhaps differences appear. But it was my definite impression that the best interfaces were achieved with the titanium alloy. If I had to choose one single material out of these four I would pick the titanium alloy. Why our results are different is a mystery I cannot explain.

Professor Solomon: Most people agree that we get better integration in those experiments where there is a screw being inserted into bone in one form or another, than with other implants, either laid on to bone or press-fitted into bone. Has the screw thread geometry anything to do with the bone ingrowth or ongrowth? I am talking about depth of thread and pitch.

Dr Linder: I see screw threads only as a means of immobilising the implant. I think that a screw is more stable in the bone than a cylindrical implant, and that osseo-integration is more easily achieved because of that. But there may be other factors too.

Professor Albrektsson: I agree, but I also believe that the screw design has an impact. The only experimental and clinical feeling I can give is that when we work with very pointed threads, we nearly always relate it to a stress concentration resorption which we can see. We rounded off the outer ridges of the screw design and achieved a much higher proportion of bone contact. That is one parameter. I think that I would like to see rounded surfaces in all kinds of dimensions on my screws. I do not want deep and very pointed threads in the screws at all.

Mr Elson: But your threads have been cut very carefully, haven't they? It is almost a sliding fit when you eventually put in the definitive screw, in the same way as Dr Linder's cylinder is also a pressed fit. There is no concept of violent friction as you put in the screw.

Professor Albrektsson: It is our aim, at least. I forgot to mention that we have been looking at various thread patterns and we have seen the advantage in the thread-to-thread distance, in that we like them to be relatively densely threaded compared to having the opposite type of pitch.

Chapter 8

Histology of the Failed Implant

A.J. Darby

Despite the current emphasis on wear debris, it should not be forgotten that infection can still play an important role in implant failure. The histological features of such an infection adjacent to an implant do not differ, however, in any significant way from similar inflammatory responses elsewhere in bone or soft tissues.

The importance of infection in the postoperative period is the effect it has on the bone—implant interface and its future structural integrity. Infection increases the extent of the necrosis already present as the result of surgical trauma. This, together with the production of inflammatory cytokines, will increase osteoclastic activity and possibly delay osteoblastic healing. The result is an increase in production of fibrous tissue preventing the desired close apposition of bone to prosthesis.

Most cases of implant failure are aseptic and it is the formation of wear debris and its contribution to prosthetic loosening that is of most interest to orthopaedic surgeons. This debris is commonly composed of three types of particulate matter: bone cement, high-density polyethylene (HDP) and metal.

Cement Debris

The cement cannot usually be seen in histological sections because it is dissolved out during tissue processing. Its absence can be recognised by the spaces it leaves and the cellular reactions it provokes (Fig. 8.1).

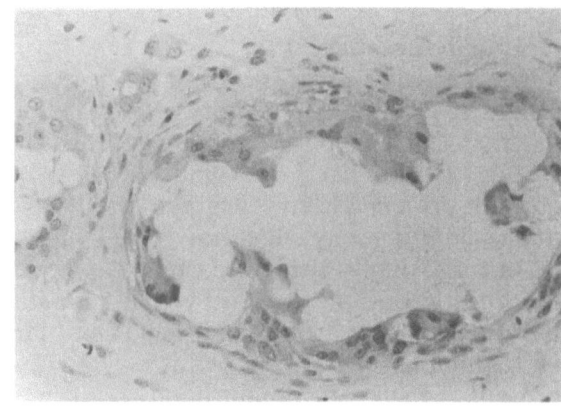

Fig. 8.1. Cement debris. Empty spaces representing dissolved cement fragments surrounded by foreign body giant cells. In some places the rounded outlines of monomer beads are clearly visible.

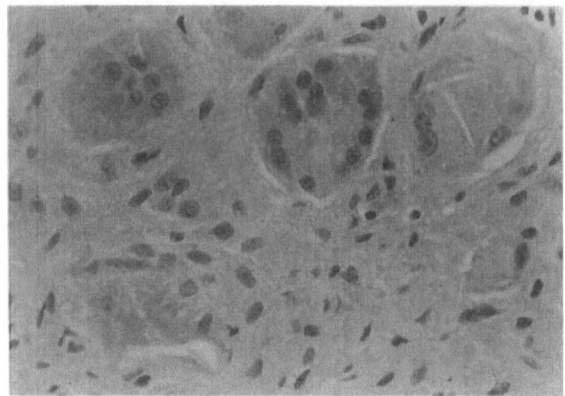

a

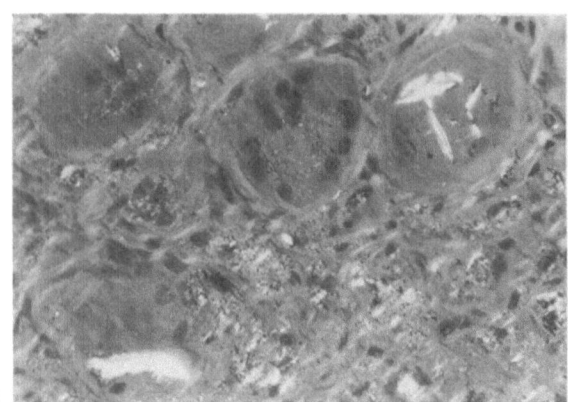

b

Fig. 8.2a,b. Polyethylene debris. **a** Giant cells and macrophages. The giant cells contain unstained elongated spaces. **b** The same field viewed under polarised light. The spaces which appeared empty in **a** are seen to contain brightly birefringent HDP debris. Smaller particles of HDP can also be seen in adjacent macrophages.

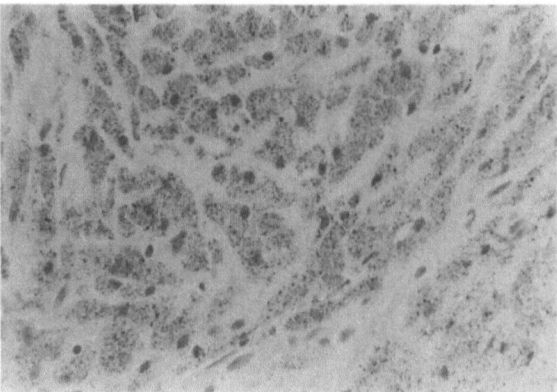

Fig. 8.3. Metallic debris. Sheets of macrophages whose cytoplasm contains numerous darkly staining particles of metallic wear debris.

Polyethylene Debris

In contrast to cement, which often seems to produce a predominantly fibrous reaction, HDP particles are characterised by a florid histiocyte and giant cell response. With traditional microscopy, HDP appears as small irregular holes or shard-like spaces and cannot be recognised (Fig. 8.2a). Unlike cement, however, it is brilliantly birefringent when viewed with polarising microscopy (Fig. 8.2b). The smaller fragments are taken up by mononuclear histiocytes, while larger pieces are present within multinucleated foreign body giant cells. All sizes of fragment may also be seen lying free in necrotic or fibrous tissue.

Metal Debris

The third type of debris, usually seen in smaller amounts than cement and HDP, is metallic in origin. Histologically, large fragments may be seen lying free in the tissues. More commonly it is apparent as small black particles within histiocytes (Fig. 8.3). Although obvious when present in large amounts, single particles may have to be searched for under high power, and identification is aided by "edge-effect" birefringence under polarised light.

Like cement and HDP, little is known about how much metal may be released into the tis-

Sometimes the barium sulphate impregnation of the cement can also be seen as a particulate residue in these spaces.

The principle tissue response to cement is progressive fibrosis together with a cellular reaction consisting of macrophages (histiocytes) and foreign body giant cells, which attempt to engulf and degrade the foreign material. Our knowledge, however, is limited by our inability to recognise cement except as the ghostly outline formed by cells moulding themselves around monomer beads and larger fragments. The sheets of histiocytes often present in the neighbourhood of cement debris may well be reacting to smaller particles that we cannot visualise. The full extent of cement debris, what it is doing and where it is going can only be a matter of speculation.

sues. Much of the metal may be of a size below the limits of resolution of light microscopy and perhaps is more easily transported by macrophages to local lymph nodes and beyond. As prostheses are being inserted into younger and younger patients, I am concerned about the possible long-term effects of increasing tissue levels of metals in distant organs. This is particularly problematic in the context of porous-coated prostheses where the surface area of the prosthesis—tissue interface is greatly increased.

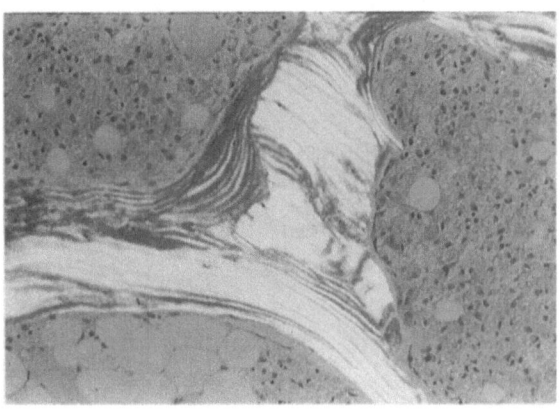

Fig. 8.4. A bony trabecula viewed under polarised light. The marrow spaces on the right and top left are filled with macrophages containing weakly birefringent debris. The right-hand side of the bone has a scalloped margin indicative of resorption and two multinucleate osteoclasts can be seen near its lower end.

The Interface Membrane

A fibrous membrane surrounds the prosthesis whether it is loose or stable. Attempts have been made to describe the structural organisation of this membrane. In practice, however, I think such accounts are highly idealised. The surface adjacent to the prosthesis or its cemented covering may vary from fibrinous exudate to granulation tissue or to synovial metaplasia within the same histological section. Elsewhere in the same membrane, surface necrosis, fibro-cartilaginous metaplasia or collagenous fibrous tissue may all be seen. Occasionally, the complete thickness of part of the "membrane" may consist of granulomatous tissue containing numerous macrophages and giant cells. In general, however, cartilage and synovium, and to a lesser extent granulation tissue, tend to be surface features, whilst the depths of the membrane adjacent to bone appear more fibrous. Wear debris also appears to concentrate in the deeper layers of the membrane. The deep margin of the membrane is by no means clearly defined. It often permeates trabeculae of reactive woven bone, and irregular areas of fibrous tissue and histiocytes, with or without wear debris, may extend into the adjacent marrow.

Synovial Metaplasia

The surface of the interface membrane often shows an ordered pattern of cells approximat-

ing to the appearance of normal synovial tissue. Some authors have postulated an important role for this in the causation of loosening. Whilst accepting that products of synovial tissue, particularly when inflamed, may stimulate osteoclastic activity and aggravate established loosening, I am not convinced of its importance in the initiation of loosening. Synovial metaplasia is not seen in the absence of loosening, and seems to occur only as a response to significant movement between the implant and adjacent soft tissues. It seems likely, therefore, to be an end-stage phenomenon.

Osteoclasts, Giant Cells and Bone Erosions

An increasingly recognised, although uncommon, complication of long-term implants is the development of massive bone erosions. These are filled with toothpaste-like material at revision surgery, and microscopy shows them to consist of fibro-histiocytic tissue, usually with extensive necrosis and variable amounts of wear debris.

The bone adjacent to such an erosion shows the marrow infiltrated by numerous histiocytes, and active resorption of bone can be recognised (Fig. 8.4). It is sometimes stated that a component of the granulomatous tissue itself can destroy bone. In fact, there is little if any evidence that macrophages or foreign

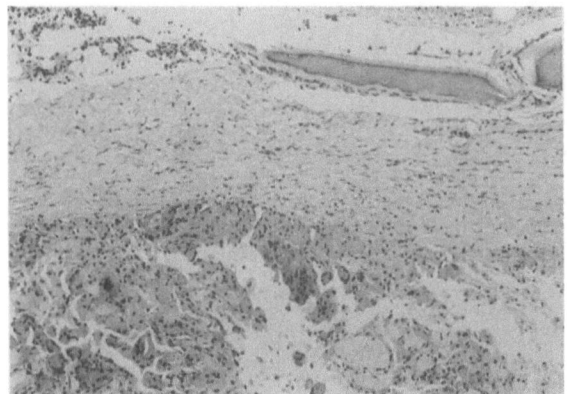

Fig. 8.5. The interface membrane from an uncemented iso-elastic prosthesis. The top part of the photomicrograph shows marrow and a bony trabecula whose surfaces are covered by osteoid tissue. Below this is a layer of pale-staining fibrous tissue and then (adjacent to the prosthesis) granulomatous tissue containing giant cells and macrophages.

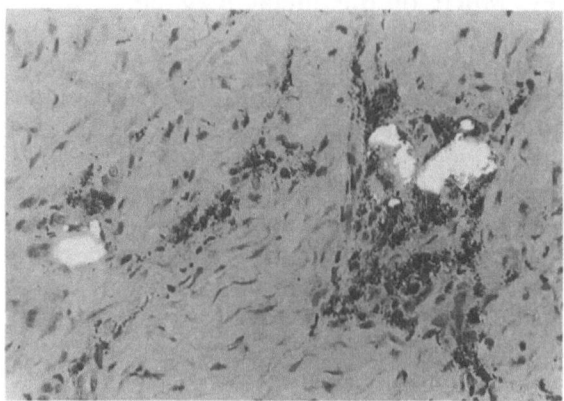

Fig. 8.6. Fibrous tissue with macrophages containing dark metallic particles. The white birefringent material is possibly derived from the polyacetyl resin of the iso-elastic stem.

body giant cells can resorb bone. Such cells, taken from the bone—cement interface, have been recently cultured on bone slices and failed to show any evidence of bone resorption (Pazzaglia and Pringle 1988). By contrast, culture of osteoclasts produced resorption pits (Howship's lacunae) similar to those seen in Fig. 8.4. It seems likely, therefore, that the cells of the interface membrane stimulate neighbouring osteoclasts to resorb bone, rather than having a direct action on bone.

Iso-elastic Stem

Recently we have begun to see a few failures of the uncemented, iso-elastic prostheses. It is interesting that the tissue reaction around these iso-elastic stems seems to differ very little from that already described with cemented prostheses. There is the same sort of granulomatous response and fibrosis, and some of the adjacent bone shows extensive surface osteoid formation (Fig. 8.5).

I do not think, however, that this osteoid is evidence of a mineralisation defect. Osteoid tissue is a normal component of bone and is seen wherever bone formation is occurring as part of the continuing process of either normal or abnormal bone remodelling. The possibility of focal osteomalacia adjacent to a prosthesis may occur, perhaps as a result of local aluminium toxicity.

The abundant metal present in some cases and the nature of the birefringent material seen was of particular interest in relation to the iso-elastic stems (Fig. 8.6). This latter would appear to differ subtly in its appearance from the more commonly seen HDP and may, in fact, be polyacetyl resin from the stem. We are awaiting results of further analysis of this material.

Does the precise nature of this foreign material really matter, however? Looking at other prostheses in other sites, anterior cruciate ligament replacements for example, particulate wear debris such as carbon fibre, polyester or Goretex always seem to produce a similar granulomatous reaction if present in sufficient quantity. In the context of a hip prosthesis, wear debris of any type will provoke an inflammatory response leading to fibrosis, osteoclastic bone resorption and eventual radiological and clinical evidence of loosening.

Conclusion

The term "biocompatible" is much favoured by both prosthesis manufacturers and users. I am not sure what it really means. Even with titanium dental implants, foreign body giant cells can sometimes be seen at the titanium—soft tissue interface. Whatever foreign material is used, it provokes a tissue reaction. In my experience, I think the problem of preventing the generation of wear particles is not so much one of biocompatibility, but depends upon the skill of the surgeon and the design of the prosthesis.

Reference and Further Reading

Pazzaglia UE, Pringle JAS (1988) The role of macrophages and giant cells in loosening of joint replacement. Arch Orthop Trauma Surg 107:20—26

Discussion

The Chairman - **Mr Elson**

The Panel - **Dr Darby**
       - **Dr Draenert**
       - **Professor Willert**

Mr Elson: You saw a great deal of metal in the failed iso-elastic cases. Is that peculiar to them?

Dr Darby: I do not know. One is always up against the problem of sampling in histology, the number of blocks one takes, the number of sections one examines. It was easy to see metal around the iso-elastic implants as opposed to many prosthetic samples where one has to look hard for isolated single metallic particles. Where all the metal originates, I do not know.

Mr Elson: Is it from the screws at the top?

Dr Darby: Presumably, but it is migrating a long way if it does come from those screws.

Mr Elson: There are other prostheses being put in now with transcortical screws of various anchoring types. Do you think the frictional effect is going to be the same?

Dr Darby: I think it may well be so.

Dr Draenert: I should like to put a question to both Dr Darby and Dr Linder. Do you think that stable contacts — bony contacts without any fibrous tissue — can be combined with micro-movement at a cross-sectional level?

Dr Darby: I hope that I did not imply that; I did not mean to do so. Certainly you will get micro-movements. I think bony trabeculae in functional continuity with the prosthesis are ideally situated for micro-movements. That is how I envisage normal skeletal functioning, with the trabeculae bending and flexing but not fracturing, with sprains.

Dr Linder: I totally agree.

Dr Draenert: There are foreign body giant cells and bony contact. Some state that foreign body giant cells are in good correlation to micro-movement which can occur. Can you imagine micro-movements along one part of a fully integrated implant?

Mr Elson: I want to be sure that we have all understood the question. You refer to a "fully integrated implant", whatever that means, and you are postulating that in different areas of this you can have micro-movement due to the elasticity effect?

Dr Draenert: Yes.

Professor Willert: I do not agree that giant cell formation is the sign of micro-movement. I cannot see the connection between micro-movement and the formation of foreign body giant cells. Some authorities feel that the amount of uncalcified osteoid is too great in relation to the other bone surrounding the implant. In the other bone there is little or no uncalcified osteoid. There is also too much uncalcified osteoid related to the time after implantation. The initial remodelling should be over in these cases of more than 5 years since implantation.

Dr Darby: Perhaps I can enlarge on this. The initial remodelling may be finished at 5 years, but there are still abnormal biomechanical stresses around some parts of that prosthesis. In such areas remodelling may well continue which will produce increased surfaces. Tetracycline labelling could quite easily be done to see whether mineralisation does occur at the osteoid surfaces. An in vivo tetracycline label is unnecessary; the tetracycline label can be given in vitro when the specimen is received in the laboratory.

Professor Willert: I wonder whether you have tried to differentiate between metal and X-ray medium. In the iso-elastic prosthesis, the tissue looked to be incorporating metal, but in the former specimen of a cemented prosthesis this could possibly be a contrast medium. We described some methods for distinguishing metal particulates from others. You do not have birefringency in metal but there is a light effect which breaks at the surface. I would recommend that you analyse the tissue for the material content. If you are not sure what you see, I would recommend that you do this for the polyester. You can destroy the soft tissues and then make an analysis of the polyester residuum.

Dr Darby: I think that in relation to the metal particles it can be difficult in some circumstances. Under polarised light, my impression is that it was definitely metal as opposed to barium. I agree that proper analysis is the only way to be certain.

I also take the point about trying to get some quantitative analysis of the tissue load of these substances. Perhaps we should also look more closely at the size distribution and the shape of particles. There is certainly evidence from other situations in the body where we have foreign material of this nature, that shape and size of particles is of paramount importance.

Ms Campbell: I should like to ask Professor Willert and Dr Draenert whether they agree with Dr Darby's observation that it is not so much the material but the presence of fine wear debris that will activate histiocytes in the response that he has seen in failed implants. If not, what is the worst material, in your opinions?

Dr Draenert: I agree with Dr Albrektsson that we did not realise there was so much difference between different materials and that the influence of design and the implant mass was so overwhelming. The body response will be most influenced biomechanically according

to your findings but there is also the biological effect.

Professor Willert: In my opinion it is not the type of material but the fact that it is released in large amounts as particulate matter that causes the granulomatous reaction. You confirmed the findings of other researchers in this respect. It seems that the principles we mentioned are as stated.

Ms Campbell: In other words, you are not really seeing a qualitative difference in the wear debris reaction between metal, polyethylene or PMMA wear debris?

Professor Willert: Yes, there are some differences. Tissue reaction to metal looks a little different from tissue reaction to polyethylene particles. The differences are brought about by the size of the particles and sometimes the chemistry. But, in general, the presence of particles which have to be phagocytised by the surrounding material is enough to badly disturb the interface and finally to cause loosening.

From a practical point of view, the different quality of the material is finally the major reason for wear and for all the problems. Of course, we are constantly looking for those materials which will cause less wear and fewer problems.

Dr Malcolm: I have been able to look at a variety of tissue reactions to prostheses in patients, particularly metal on metal, which has naturally produced metal wear particles. Also ceramic particles where there have been revisions, and of course large amounts of high-density polyethylene. I suspect it is the size and quantity of the particles produced that are most relevant to tissue reaction. The type of reaction that I have seen to ceramic particles is quite different, however, from that to high-density polyethylene. The chemical nature probably plays a lesser role.

Dr Darby: I have no experience of ceramic particles but I think I would agree in broad terms.

Dr Clarke: Dr Darby, you showed millions of histiocytes invading the bone, and yet you clearly emphasised that it was the osteoclast doing the damage. Is there no way that the histiocyte and the macrophage can be equally dangerous?

Dr Darby: I have not read of it. There are papers in the literature claiming that histiocytes and macrophages can resorb bone, but those experiments are not what I would call biological experiments in the sense of looking at histiocytes to resorb bone. They are experiments where histiocytes are incubated with bone chips, for example, and calcium is released. I do not think we have an experimental analogy with the in vivo resorption of bone that we know occurs in normal bone remodelling.

Mr Atkins: You showed us one failed femoral component where there was massive lysis and endosteal resorption. Was there a very active osteoclastic resorption front microscopically?

Dr Darby: This is a paradoxical situation. Very rarely are large numbers of osteoclasts seen. Osteoclasts are short-lived cells which migrate actively along the bone surface. You can have a variant of very active resorption and yet see very few osteoclasts in a single histological section. All we have histologically is the evidence of the scalloped margin, the Howship's lacunae. I cannot really answer your question positively, but I still believe that it is osteoclasts that do the damage.

Mr Elson: I am struggling to keep up with all this high frontier knowledge. As I understood the position, when you inserted cement there was always a zone of necrosis, and that necrotic bone has to be absorbed by something. Is that absorption by osteoclasts followed by remodelling and then areas of osseo-integration?

Dr Draenert: Yes, it is certainly resorption by osteoclasts. It can be clearly shown by scanning the Howship's lacunae.

Mr Elson: Subsequently new bone is then laid down, which comes into contact?

Dr Draenert: Yes.

Mr Elson: Thank you. That is wonderful. I understood that!

Chapter 9

The Effects of Wear Particles on Periprosthetic Tissues

D.W. Howie

Circumstantial evidence exists which implicates wear particles in bone resorption. We have performed in vivo studies to substantiate the hypothesis that bone resorption may be stimulated by the macrophage and multinuclear giant cell response to wear particles (Fig. 9.1) (Willert and Semlitsch 1977). It was commonly thought, until 5—7 years ago, that the bone resorption commonly seen around joint replacement prostheses was usually due to low-grade infection.

Clinical Experience

It is common to observe examples of excessive prosthetic wear particle production, such as in the case of some metal-on-metal prostheses, where the bone resorption that these implants occasion can be excessive (Fig. 9.2). Histological examination of the periprosthetic tissues revealed a response to very fine metal particles and even around stable implants granulation tissue may be seen invading the bone.

The extent of metal deposition can be seen more clearly using electron microscopy with-

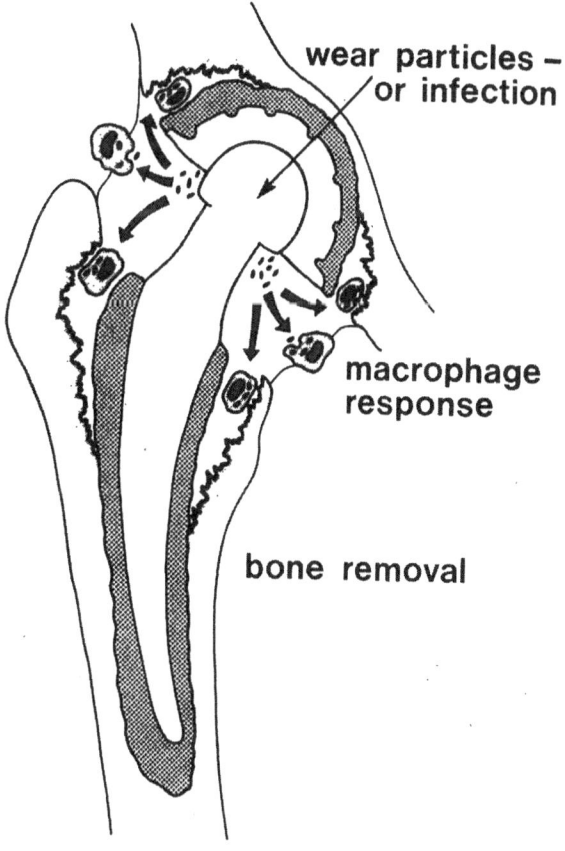

Fig. 9.1. Bone resorption may be stimulated by macrophage and multinuclear giant cell response to wear particles.

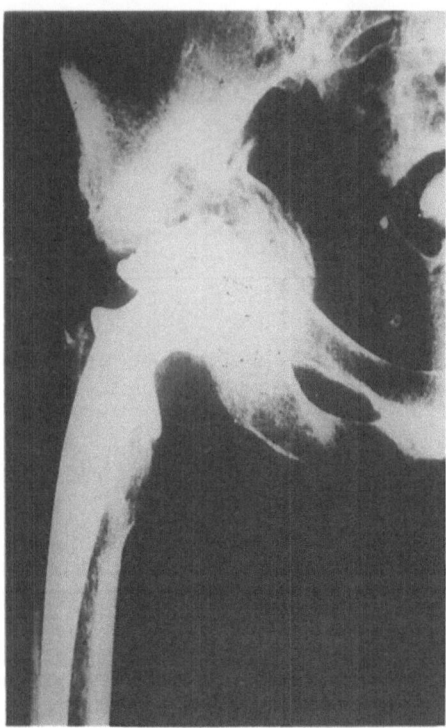

Fig. 9.2. Radiograph of metal-on-metal hip prosthesis associated with excessive bone resorption.

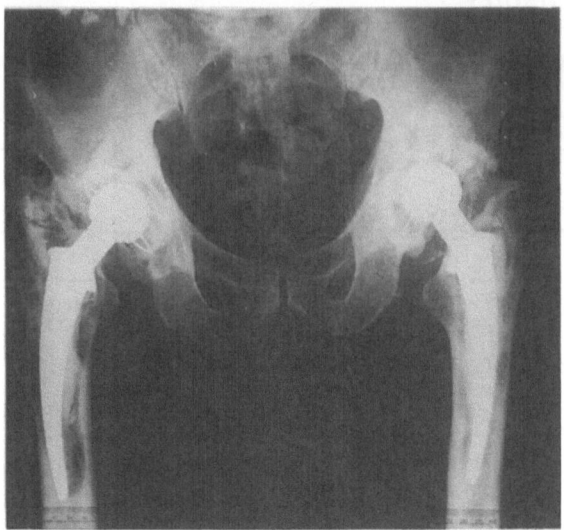

Fig. 9.3. Radiograph of bilateral metal-on-polyethylene prosthesis associated with bone lysis.

bone. Cement had remained impregnated between the bone trabeculae. The cement was in close contact with the bone in that region, yet large amounts of wear debris were

out which the very fine particles within macrophages are not appreciated.

Metal-on-polyethylene prostheses are also associated with this bone lysis (Fig. 9.3). The findings are of considerable interest when an implant is revised (Fig. 9.4). In the case shown in Fig. 9.4 there was a small area of lysis near the tip of the stable component, but little evidence of wear of the polyethylene acetabular components. Examination of the tissue from the bone—cement interface at the tip of this apparently stable prosthesis showed macrophages containing fine polyethylene wear particles (Fig. 9.5).

We examined the tissue around hip arthroplasties requiring revision surgery. These arthroplasties were a human model of excessive wear particle production. Important examples were those cases where there was no appreciable loosening of the femoral component. Examination of the apical zones revealed a difference compared to the basal zones. In the basal zones there was connective tissue at the bone—cement interface while in the apical zones there was evidence of intimate contact between cement and

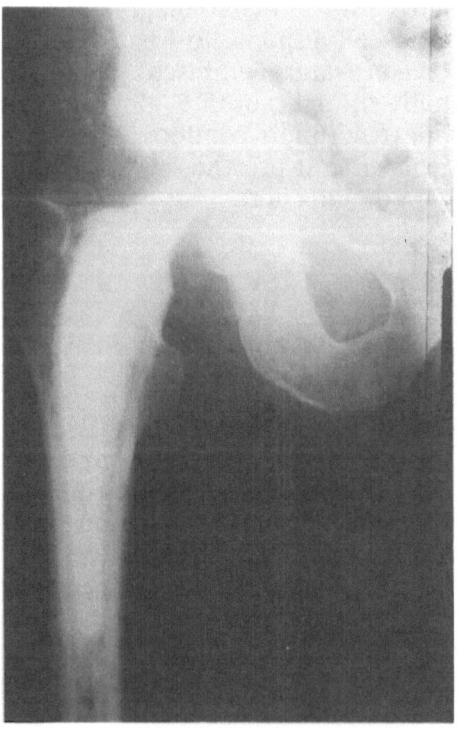

Fig. 9.4. Radiograph showing small area of lysis around tip of femoral component.

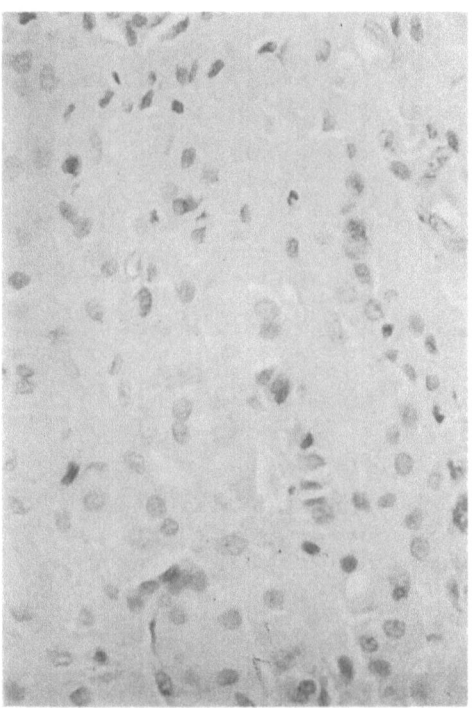

Fig. 9.5. Photomicrograph of tissue from bone—cement interface around tip of femoral component shown in Fig. 9.4 shows macrophages containing fine polyethylene wear particles.

seen at the interface more than 2 cm from the edge of the component. These findings demonstrated that fine particulate debris can migrate along apparently stable interfaces.

Ceramic-on-ceramic prostheses, of which we have had little experience, will produce wear particles, particularly if the components are malpositioned. Finally, it is important to emphasise that when any material abrades against bone, wear particles will be produced at this interface. The number of particles will depend, among other things, on the hardness of the implant material. From the observations described above there is good circumstantial evidence that prosthetic wear particles produce an adverse response in the periprosthetic tissues and this response is associated with bone resorption.

Experimental Work

As well as human studies we undertook in vivo and in vitro studies to try to piece together the problems associated with prosthetic wear particles (Fig. 9.6). In cell culture it is known that metal particles will occasion a rounding up and necrosis of macrophages, the degree of necrosis depending on the amount of cobalt.

The synovial and subsynovial response to injection of chrome—cobalt particles into rat knee joints is shown in Fig. 9.7 (Howie and Vernon-Roberts 1988c). The significance of a peak of lymphocytes at 7 days was unclear. At 1 week, electron microscopy confirmed the large number of particles and the loss of the normal synovial layer. In other areas, particles were seen to be contained in macrophages. At 3 months, macrophages and necrosis persisted in the tissues.

Electron microscopy confirmed the small number of particles and the loss of the normal synovial layer. In other areas, particles were seen to be contained in macrophages.

In further studies of the long-term effects of chrome—cobalt particles, it was found that after 6 months and 1 year necrotic villi may persist with particles lying within the amorphous ground substance (Howie and Vernon-Roberts 1988a). We mapped the particle score (Fig. 9.8) and the macrophage score in the subsynovium over 1 year (Fig. 9.9). While some particles were cleared from the joint, others remained within the restrictions of this measurement technique (Howie and Vernon-Roberts 1988a). Particles tended to remain distributed around the joint with the associated macrophage response persisting within the joint. So while particles are being cleared, if there is continued production of wear particles by a joint prosthesis they will still remain and accumulate within the joint.

The effect on tissues of aluminium oxide particles was compared to that of cobalt—chrome particles of similar size (Howie and Vernon-Roberts 1988b). One week following injection there was a difference in the macrophage to particle response, suggesting that the response to the cobalt—chrome in terms of the number of macrophages was more severe. At 4 weeks and at 3 months we did not find a difference in the macrophage response (Fig. 9.10). The type of wear particle may, therefore, determine the tissue response.

Polyethylene particles were prepared in a wear simulator. Injection of these particles

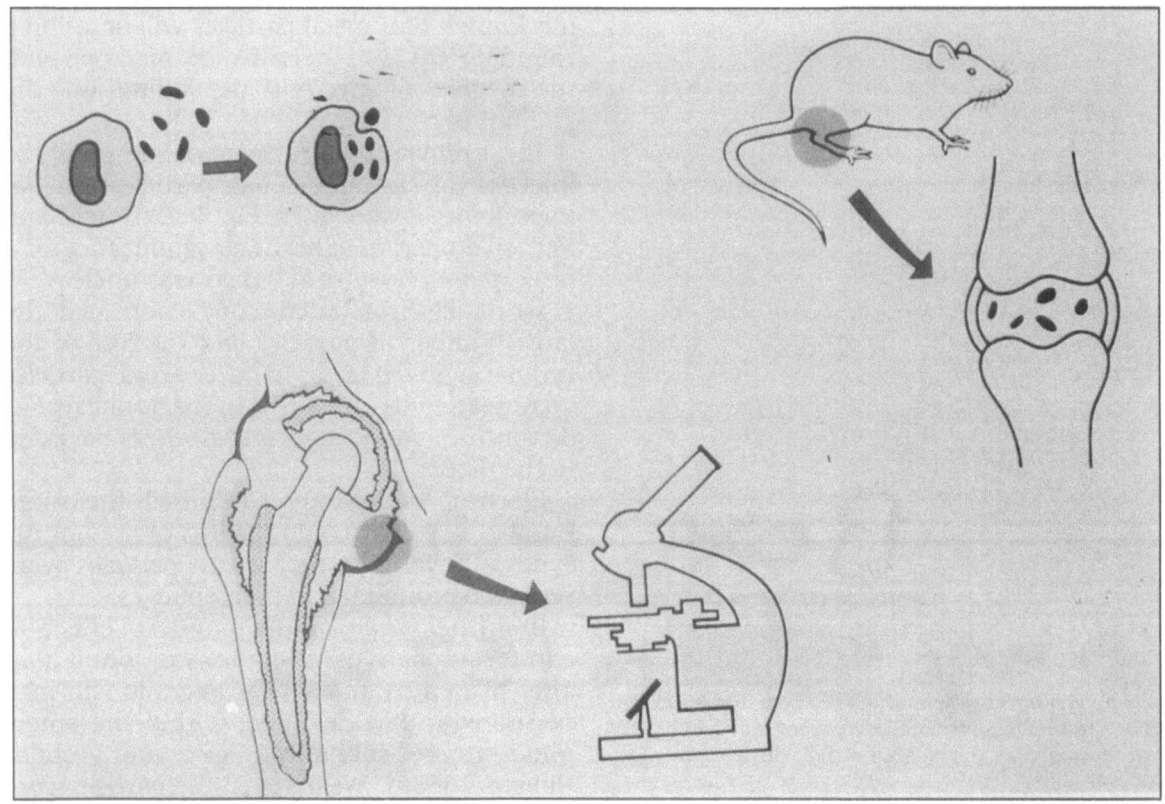

Fig. 9.6. Diagrammatic scheme to show in vivo and in vitro human and animal studies to analyse the problems associated with prosthetic wear particles.

produced the typical appearance seen around a polyethylene prosthesis. Large shards of polyethylene and also aggregates of small particles were contained within multinuclear

giant cells. Examination of tissue revealed that once a particle reaches the size of a macrophage nucleus that particle tends to stimulate a multinucleate giant cell response

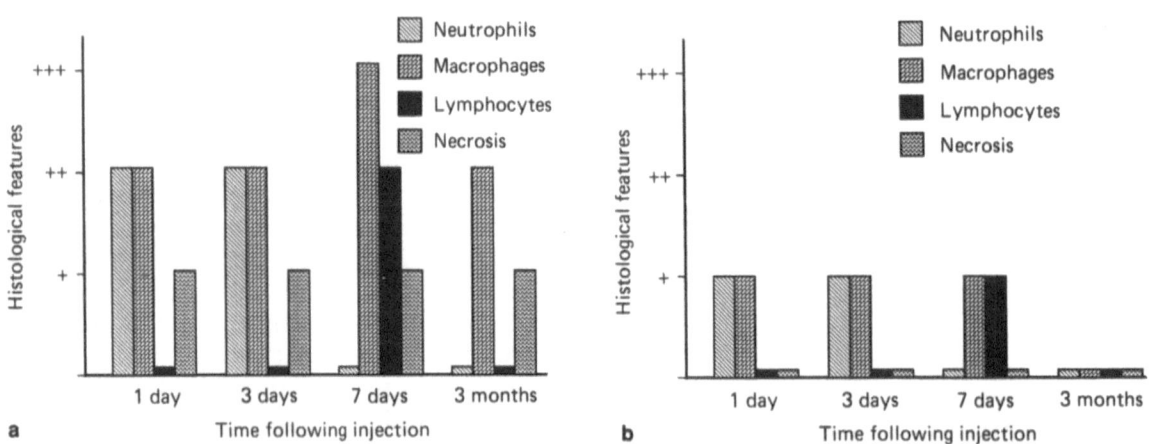

Fig. 9.7a,b. Histograms showing **a** the histological features of cobalt—chrome particles and **b** saline control in rats at various time periods following injection. Features are graded as 1+ (occasionally present), 2+ (present in up to half the knee) or 3+ (present in more than half the knee) (Howie and Vernon-Roberts 1988c).

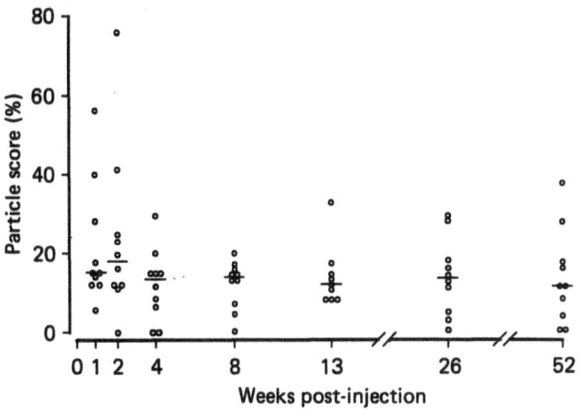

Fig. 9.8. Median and individual particle scores of knees at various time intervals following injection of cobalt—chrome particles.

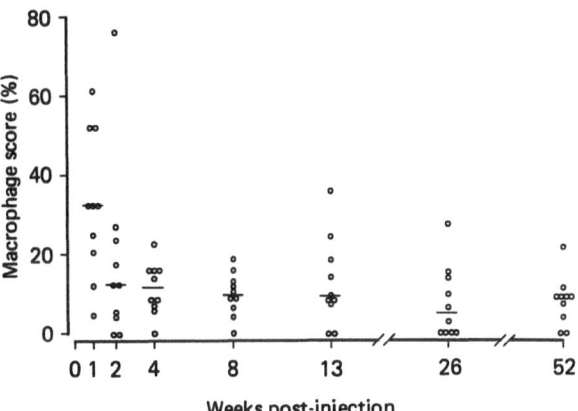

Fig. 9.9. Median and individual macrophage scores of knees at various time intervals following injection of cobalt—chrome particles.

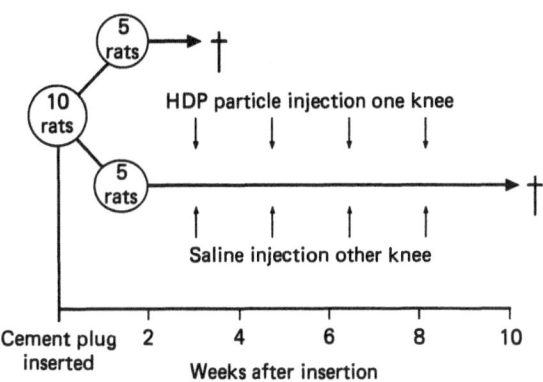

Fig. 9.11. HDP particle injection.

rather than a mononuclear cell response.

Finally, we inserted an acrylic plug into the distal femur of rats and then injected polyethylene particles into the adjacent joint (Howie et al. 1988). Ten rats had plugs inserted and we sacrificed five of them at 2 weeks. In the other five, after 2 weeks we began repeated injections of polyethylene wear particles in one knee and saline particles in the other knee (Fig. 9.11). If an acrylic plug is inserted into the femur of a rat, the plug rapidly becomes encased in bone (Fig. 9.12). The cement is closely applied to the bone. If one injects saline there is no change in that appearance, whereas repeated injections of polyethylene particles produce a different appearance. Polyethylene particles were found within the knee joint, but also at the interface between the cement plug and bone. A transverse section of the bone near the

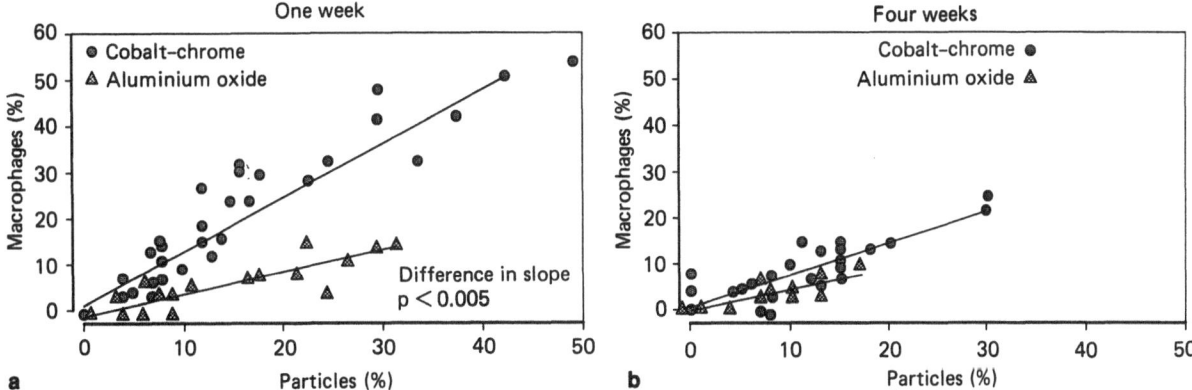

Fig. 9.10a,b. The macrophage response at 1 (**a**) and 4 (**b**) weeks following injection of Co–Cr and Al_2O_3 particle suspensions. The slopes are significantly different at 1 week but not at 4 weeks following injection, indicating a more severe macrophage response to similar numbers of Co—Cr particles compared to Al_2O_3 particles.

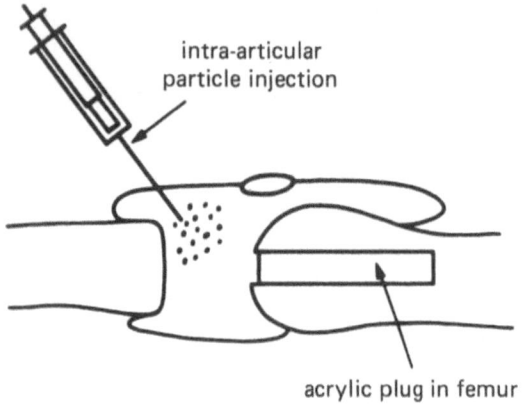

Fig. 9.12. The site of injection of particles into a knee joint, adjacent to a previously inserted plug of acrylic cement (from Howie et al. 1988, with permission).

joint showed that instead of the plug being encompassed in bone there was now connective tissue with voids which contained the polyethylene particles.

Conclusions

We feel this model confirmed our hypothesis that in the absence of infection and loading and, therefore, any significant movement at the interface, one sees bone resorption in response to prosthetic wear particles (Fig. 9.13).

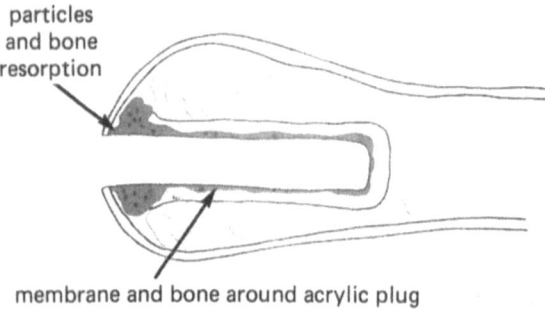

Fig. 9.13. Diagrammatic representation of a sagittal section of the femur of a rat, showing resorption of bone and the formation of a connective tissue membrane at the interface between the acrylic cement and the bone in response to injection of particles of polyethylene into the adjacent joint (from Howie et al. 1988, with permission).

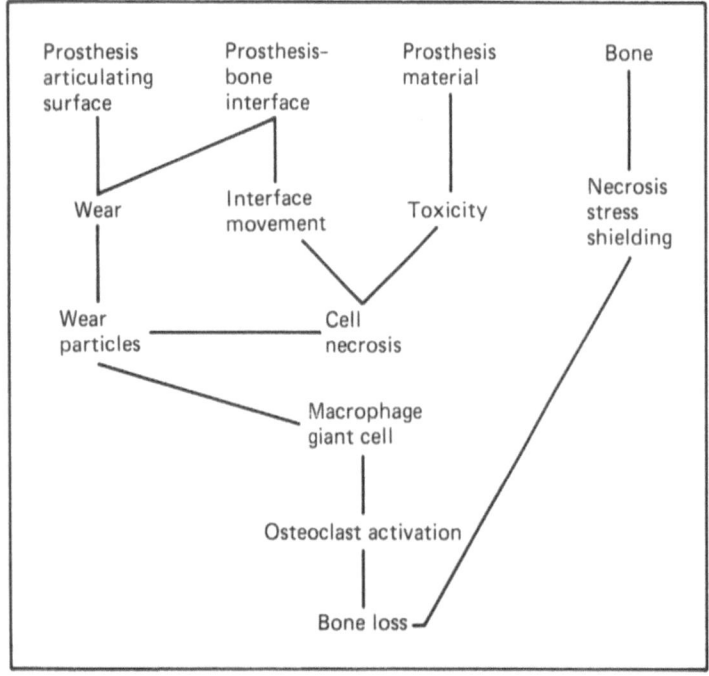

Fig. 9.14. Causes of aseptic loosening.

Summary

The biological causes of loosening and the effects of wear particles are summarised in Fig. 9.14. The articulating surfaces of prostheses will produce wear particles which induce a macrophage and giant cell response, and so stimulate bone resorption. Movement at the prosthesis—bone interface may produce wear particles, which induces bone loss through the same mechanism. Obviously, toxicity of the material and other factors such as necrosis of bone and stress shielding will also contribute to bone loss around an implant.

Acknowledgement. Professor Howie is indebted to Professor B. Vernon-Roberts for his assistance in the preparation of this material.

References and Further Reading

Howie DW, Vernon-Roberts B (1988a) Long-term effects of intra-articular cobalt—chrome alloy wear particles in rats. J Arthroplasty 3:327—336

Howie DW, Vernon-Roberts B (1988b) Synovial macrophage response to aluminium oxide ceramic and cobalt—chrome alloy wear particles in rats. Biomaterials 9:442—448

Howie DW, Vernon-Roberts B (1988c) The synovial response to intra-articular cobalt—chrome wear particles. Clin Orthop 232:244—254

Howie DW, Vernon-Roberts B, Oakeshott R, Manthey B (1988) A rat model of resorption of bone at the cement—bone interface in the presence of polyethylene wear particles. J Bone Joint Surg (Am) 2:257—263

Vernon-Roberts B, Freeman MAR (1977) The tissue response to total joint replacement prostheses. In: Swanson SAV, Freeman PAR (eds) The scientific basis of joint replacement. Pitman Medical, Tunbridge Wells, pp 86—129

Willert HG, Semlitsch M (1977) Reactions of the articular capsule to wear products of artificial joint prosthesis. J Biomed Mater Res 11:157—164

Chapter 10

Failure Modes of Artificial Joint Implants Due to Particulate Implant Material

H.G. Willert

In the early days of the low-friction arthroplasty we tried to improve the wear characteristics of Teflon and used a micro-filled Teflon for the socket. With this material, large amounts of plastic particles were abraded which produced a granulomatous reaction in the surrounding tissues of the joint capsule and bone. First indications of implant failure were seen in radiological examinations at yearly follow-up. There was progressive bone resorption at the calcar and migration of the head into the plastic socket. Pain and instability of the implant were the reasons for revision surgery.

In the process of wear, countless particles were released into the joint cavity. The reactive cell layers on the surface of the newly formed capsule subsequently phagocytosed the particles forming a granulation tissue which was predominantly characterised by persisting foreign body reaction. The spread of the granulation tissue transformed the articular capsule into a mass of fibrous tissue not only surrounding the artificial joint but also extending towards the ligaments and muscles. Within this fibrous scar tissue, large areas were poorly vascularised and became necrotic.

At revision, the inner surfaces of the capsule were found to be covered with paste-like necrotic debris. Histology of these tissues revealed particulate plastic material incorporated in macrophages and giant cells, forming large foreign body granulomas.

Initially, we hoped that the granulomatous reaction causing bone resorption around the prosthesis was a unique feature of Teflon, but we soon realised that, in principle, particles of every other implant material can also cause lysis at the bone cement interface if they are released in amounts which exceed a certain limit.

Polyester (polyethylene terephthalate) was used in a prosthesis which consisted of a metal cup, a stem with a trunnion bearing and a rotating polyester head. Its convex surface caused considerable wear so that the same storage mechanism within the surrounding capsule and at the cement–bone interfaces was induced (Fig. 10.1).

On the basis of these findings we evolved the following thesis, which we put forward for the first time in 1974 at a combined meeting of the Dutch and Swiss Orthopaedic Societies. The tissues of the joint capsule not only store the particles but obviously seem able to transport them away. This is thought to happen in the perivascular lymph spaces. This mechanism can not only eliminate wear products to a certain extent, but is also responsible for the involvement of tissues further away from the place of particle origin.

The fact that the majority of artificial joint replacements functioned very well over a

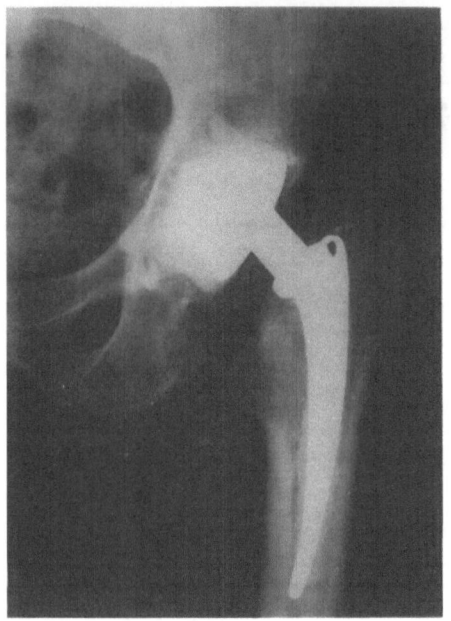 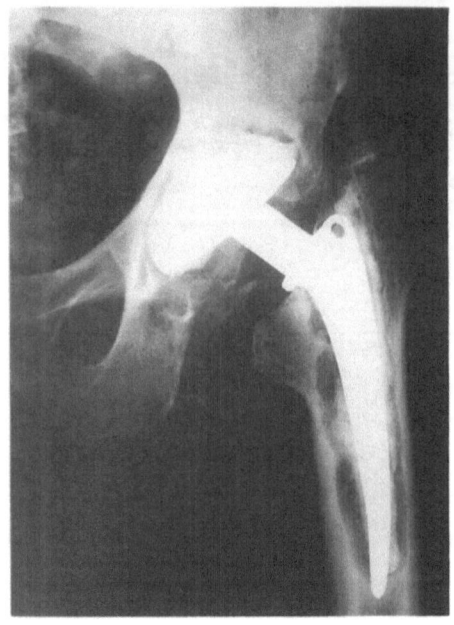

a   b

Fig. 10.1a,b. Massive scalloping osteolysis in the femur due to polyester wear. Patient with a trunnion-bearing total hip replacement consisting of a metal stem with a pivot, a polyester ball head and metal acetabular implant. Antero-posterior radiographs taken at regular follow-up. **a** One year after insertion of the prosthesis, radiolucent lines in the femur separate the bone cement from bone. **b** Massive osteo-lysis with scalloping excavations in the cortical bone of the femur and radiolucency in the acetabulum are responsible for instability and pain. The prosthesis had to be revised 2 years after implantation.

long period suggests that an equilibrium between wear and tissue reaction must be established. But when greater amounts of wear products are released, the foreign body reaction and transport activities of the joint capsule can no longer compensate and the granulomas grow to a size where they become necrotic. The wear products then find their way into other tissues surrounding the implant, in particular into the bone marrow and the bone–cement interface where they evoke a foreign body reaction. Since the neighbouring bone represents a barrier against the formation of granulomas, it is resorbed by osteoclasts. The spread of the osteolysis is then responsible for instability, local overload and finally loosening of the entire implant.

This mechanism has been found by others with the following materials of joint articulations: polymethylmethacrylate (Plexiglas), polytetrafluorethylene (Teflon), polyester, polyoxymethylene (polyacetal, Delrin), polyethylene (RCH-1000), silicone rubber, iron- and cobalt-based metal alloys, aluminium oxide ceramic (Biolox, Frialit) and carbon-fibre-reinforced polyethylene or carbon.

Our results and conclusions were confirmed by other investigators. Nevertheless, others believe that articular wear plays no part in initiating the loosening process. They assume instead that mechanical instability and disruption of bone cement anchorage is the primary cause of bone resorption. In their opinion, foreign body granulomas only aggravate the instability as a reaction to wear. We are able to show that both greater amounts of wear particles and shattering of the cement fixation may solely cause osteolysis.

Massive Osteolysis Due to Polyethylene Wear

The effects of massive polymer wear may be demonstrated by metal-on-polyethylene total joint replacements where the metal head has been scratched, e.g. by contrast media abraded from the overlapping acetabular cement cuff. Even more particles are released by convex rotation ball heads made out of a poly-

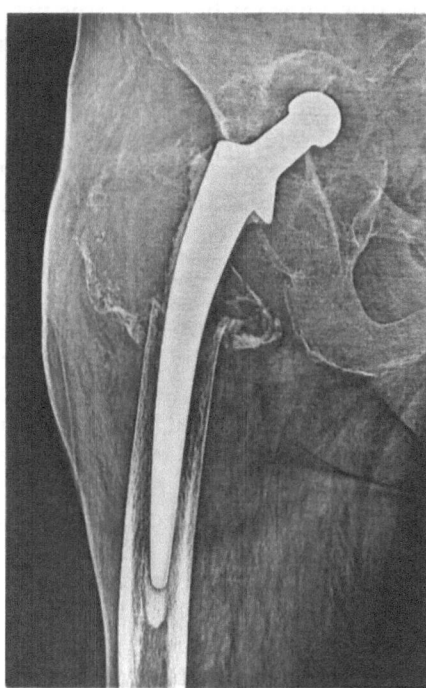

Fig. 10.2. Tumour-like osteolysis around a "soft-top" endoprosthesis. Xero-radiograph of a "soft-top" endoprosthesis 4¹/₂ years after implantation. The polyethylene rotating ball head on a 22-mm diameter ball head of the femoral component is invisible. The entire cement cuff is well preserved and its distal portion is firmly anchored in the cortical bone. A thin line of sclerotic bone surrounds a huge mass of polyethylene-storing granulation tissue.

meric material such as polyester or polyethylene.

As an example, a patient who sustained a femoral neck fracture was treated with a femoral replacement consisting of a femoral component with a 22-mm ball head with a large polyethylene head riding on the femoral stem and articulating against the natural acetabular cartilage. Five years later she was referred to us with a suspected tumour since she had developed a large osteolysis with outward bulging of a thin bone shell in the intertrochanteric region. In spite of the proximal lytic destruction, the prosthetic stem was still firmly anchored. A similar osteolysis with bulging and scalloping but considerable thinning of the bone had developed in the pelvis (Fig. 10.2). The soft-top polyethylene head was destroyed by surface abrasion of the polyethylene.

A breast carcinoma was reported in her history, but the histological tissue examination only revealed massive granulomas rich in polymer wear storing histiocytes, foreign body giant cells, fibrosis and osteoclast activity at the bordering bone. Only polyethylene was stored without any signs of polymethylmethacrylate.

The granulomas at the bone cement interface had an enormous capacity to resorb the bone (Fig. 10.3, opposite p. 72). This tissue had migrated into and widened the channels perforating the cortical bone and spread under the periosteal surface.

Osteolysis Due to Bone Cement

In the last few years, special attention has been paid to the preparation of the implant bed as well as to the handling and application techniques of bone cement. Still we have to re-operate on patients whose implants have been anchored with an inferior bone cement cuff due to incorporation of air, laminations with blood or separated portions. These implants are prone to mechanical failure of the cement implant with subsequent fragmentation and wear of cement against bone and implant against cement.

A patient received his prosthesis for osteoarthritis and was able to use it without pain for 9 years. Radiographs revealed a progressive osteolysis in the femoral shaft. Consecutively he developed pain and 10 years post-operatively the osteolysis had led to instability (Fig. 10.4). The hip was revised and massive amounts of granulation tissue between implant and bone were found. Histology, using frozen and paraffin sections, showed that mainly polymethylmethacrylate clusters, pearls with matrix and fragments thereof were stored. The larger ones were surrounded by foreign body giant cells and smaller fragments were stored in macrophages. We also found cells loaded with zirconium dioxide (Fig. 10.5, opposite p. 72).

There were clear signs of a mechanically induced separation of the cement implants due to displacement of implant relative to the cement and both relative to the bone. The surfaces of cement cuff fragments often showed highly polished areas in the contact planes where particulate material had been worn off and zirconium dioxide had been released. Fracture lines in thinner parts of the cement anchorage showed the onset of parts being broken apart. Experimentally loaded bone cement showed that the bone cement may fatigue after a certain amount of cyclic loading. The cracks produced either originate from the bone cement interface or from around the spheres, and then protrude into the other mass of polymethylmethacrylate. These results demonstrate that mechanical overload can destroy the cement independent of the location and that it might very well start right around the tip of the prosthetic stem.

We are not sure whether bone cement can be destroyed by active cellular digestion or to a significant degree by degradation with body fluids. In my opinion there is not yet very conclusive proof, only the possibility that the cement is weakened and consequently breaks into pieces. Then again the X-ray contrast medium can be phagocytosed by the surrounding tissues.

An example of local destruction of the cement cuff was seen in a patient who received a hip replacement which was in use for 6 years with no complaints. She started to complain of increasing pain and we saw a little loosened zone around the tip of the stem. The next year she reported with a greater swelling, and suspecting a tumour on the basis of a breast cancer in her history, we proposed complete revision surgery (Fig. 10.6). The patient refused to have a revision but agreed to local inspection. A tumour was found and when the tumourous mass was removed the metal stem became visible. There was neither cortical bone nor bone cement left in the surrounding area. Histology of the bone—cement interface showed no malignancy but massive granulomatous reactions, and again this granulomatous tissue protruded into the Haversian channels

resorbing the adjacent bone. In this tissue we found not only polymethylmethacrylate inclusions but also polyethylene in great amounts, but the polyethylene particles were concentrated in a few bands.

We have defined a second stage of osteolysis which is originally induced at the bone—cement interface by shattered bone cement, but is enhanced by cement fragments entering the joint space and disturbing the gliding surfaces. This considerably increases the production of polyethylene wear, and the accelerated loss of material represented by the migration of the metal head into the polyethylene socket can be seen in radiographs. Thus polyethylene particles are transported into those tissues already involved in a process of particle storage, i.e. the joint capsule and soft tissues of the acetabular and femoral bone—bone cement borders. At that time it is no longer possible to decide from histology where the process of particle production and osteolysis induction started.

In a stable implant there may be some micro-movement with a small synovial-like membrane. Some bone resorption may occur (either within the normal transformation rate of bone or by mechanical or chemical irritation), but it is usually compensated for by new bone formation to maintain an equilibrium between applied load and strength of tissue. If osteolysis occurs for any reason such as infection, insufficient mineralisation, fragmentation of bone cement or deposition of wear debris, it can start a chain reaction. The actual time for the onset of this process depends on various factors; as yet it cannot be avoided and at best can only be minimised by the careful selection of implant materials, the skill of the surgeon and moderate loading by the patient.

The use of cementless implants will help to minimise the problems of implant loosening once an appropriate design has been found. Although other problems arose, these un-

⟶

Fig. 10.4a–f. Shattering of femoral bone cement cuff at the lower end of the prosthetic stem. Series of radiographs of a patient's left hip taken at follow-up examinations; 10 years of function of a cemented total hip replacement with Co–Cr alloy femoral and polyethylene acetabular components. **a** Antero-posterior view; good, painless function 1 year after implantation, some ectopic bone appears. **b** Antero-posterior view $7^1/_2$ years post-operatively; the tip of the stem is not perfectly surrounded by bone cement, early osteolysis in this region without clinical symptoms. **c** Antero-posterior view 9 years postoperatively; onset of pain, osteolysis is limited to the area already seen before, no radiolucency in the calcar region, more severe ectopic bone formation. **d** Axial view of **c** demonstrates thinning of cortical bone by resorption. **e** Antero-posterior view prior to revision surgery; osteolysis of the lateral cortical bone becomes obvious. **f** Axial view of **e**. The defect in cortical bone has widened; localised perforation by total loss of bone seems possible.

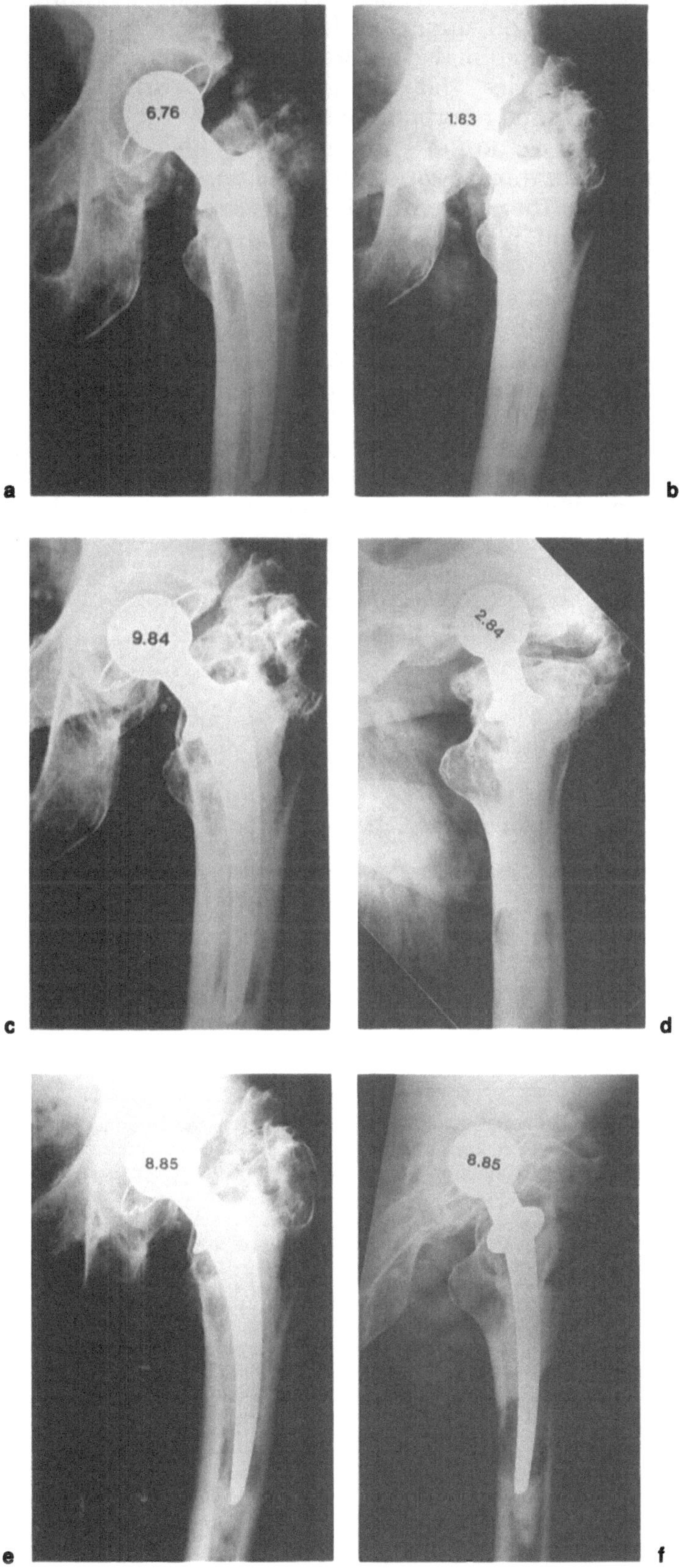

a

b

c

d

e

f

Fig. 10.4

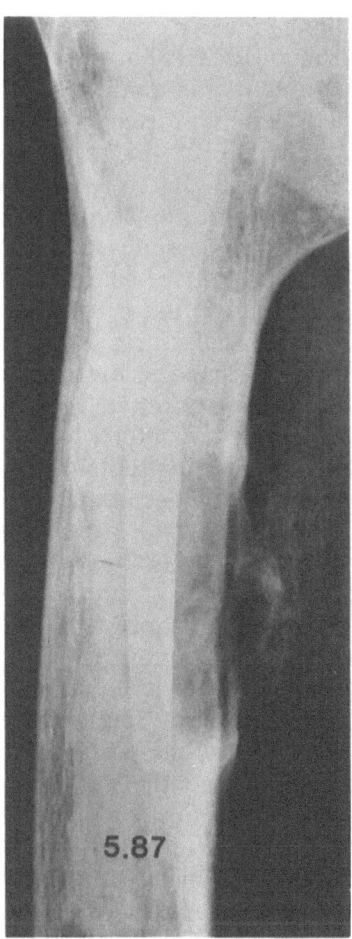

Fig. 10.6. Shattering of bone cement causes complete bone resorption. Radiograph, tangential view 7 years 7 months post-operatively of a cemented total hip replacement. The lower portion of the cemented Co–Cr metal shaft with defective cement anchorage and complete loss of cortical bone is shown. The defect is covered subperiostally with some newly formed bone.

cemented implants were thought not to have problems of wear particles. In spite of this, we still have to face a problem with an all-polyethylene cementless screw-type socket. Polyethylene was used in a very unfavourable design which, at first sight, had nothing in common with the convex spherical shape of "soft-top" joint replacements, but which, in the long run, in spite of the conical threaded outer shape, caused comparable trouble.

The initial period 4 to 6 years after operation was satisfactory, with bone formation indicated by sclerosis around the thread of the polyethylene socket in the acetabulum (Fig. 10.7). However, afterwards bone resorption started in some patients causing an extensive widening of the acetabular cavity. The sclerosis disappeared, the acetabular cup became loose and pain was the major reason for revision surgery.

We found not only thickening of the joint capsule but also massive storage of polyethylene wear from the outer surface of the polyethylene cup. This occurred more in the superior part of the acetabulum than the inferior part where there was more connective tissue. Newly formed bone came into close contact with the polyethylene by encroachment of bone spikes on to its surface. It obviously abraded the polyethylene surfaces and produced particles which were then released into the surrounding area causing bone-resorbing granulation tissue. Due to transportation of wear particles, there was osteolysis not only around the socket but also around the titanium stem in the femur (Fig.

10.8). The tissue that we removed during revision surgery of femoral cementless stems showed massive storage of polyethylene wear particles and the granulation tissue again caused resorption of bone around the stem.

We do not know the mechanisms of particle transportation, but one possibility at least is that it occurs via the perivascular vessels and lymph spaces. We have overcome this problem using sockets of the same design but with a protective titanium metal backing. This is thought to significantly reduce the risk of implant loosening by wear as there is only a minimal amount of wear originating from the articulation of the alumina ceramic ball head against the polyethylene bearing. As this amount should allow for a long-lasting equilibrium between the production of particles and elimination by transportation, we hope that the time of pain-free function of cemented joint replacements will be surpassed by the use of cementless anchored implants.

Discussion

The Chairman - **Dr Linder**

The Panel - **Professor Howie**
          - Professor Willert

Professor Fitzgerald: Lee Riley suggests that the macrophage is working as two different types; one that releases enzymatic material for the breakdown of osseous tissue, the other

---

Fig. 10.3a,b. Granulation tissue induces osteoclasts for resorption of cortical bone. Histology taken from the proximal femur during revision surgery of the patient shown in Fig. 10.2. **a** The invading foreign body granulation tissue induces osteoclasts to resorb the cortical bone. Paraffin embedding, Goldener stain, bright light, original magnification ×125. **b** Relatively large polyethylene wear particles have reached this area and are to be found in obese histiocytes and foreign body giant cells. The cells involved in the resorption of bone do not store wear particles. The typical structures of bone cement store wear particles. The typical structures of bone cement fragments are absent. Same tissue section and magnification as in **a** in polarised light. Note birefringency of polyethylene wear particles.

Fig. 10.5a,b. Granulation tissue from the site of osteolysis in the cortical bone. Histology taken from the proximal femur during revision surgery of the patient shown in Fig. 10.4. **a** Granulation tissue contains histiocytes and foreign body giant cells which store particles of a shattered cement cuff. Larger particles are surrounded by giant cells (empty spaces represent dissolved bone cement), smaller ones are incorporated either in histiocytes or in giant cells. The cells also store small crystalline, yellowish granules. Paraffin embedding, H & E stain, bright light, original magnification ×400. **b** Same tissue section and magnification as in **a** in polarised light. Note birefringency of the tiny particles which may be the liberated contrast media.

Fig. 10.7a,b. Bone within the thread is replaced by granulation tissue. Histology taken from the anchorage of a screw-in socket at revision surgery. **a** A foreign body granulation tissue has filled the space within the polyethylene threads (empty space in upper left corner); it mainly consists of histiocytes and foreign body giant cells in bundles of fibrous tissue. Paraffin embedding, H & E stain, bright light, original magnification ×50. **b** Same tissue section and magnification as in **a** in polarised light. Note birefringency of the polyethylene particles. Larger particles are surrounded by giant cells, smaller ones are predominantly incorporated in histiocytes.

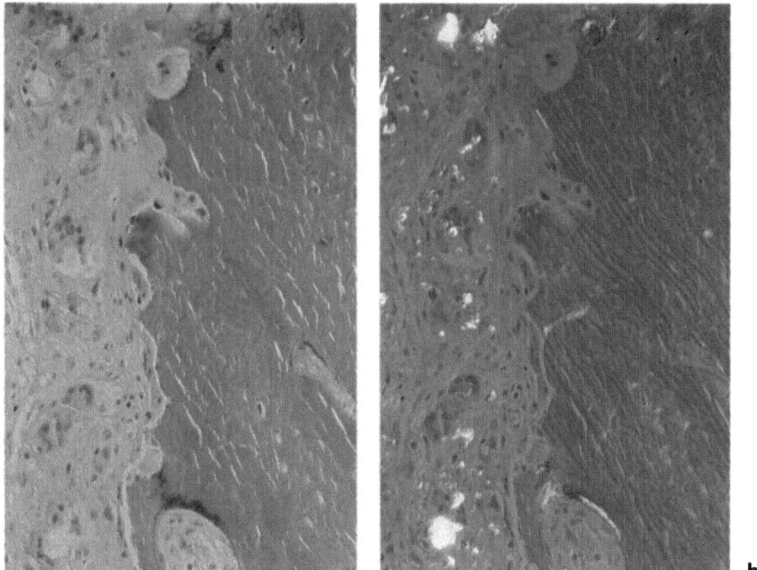

Fig. 10.3

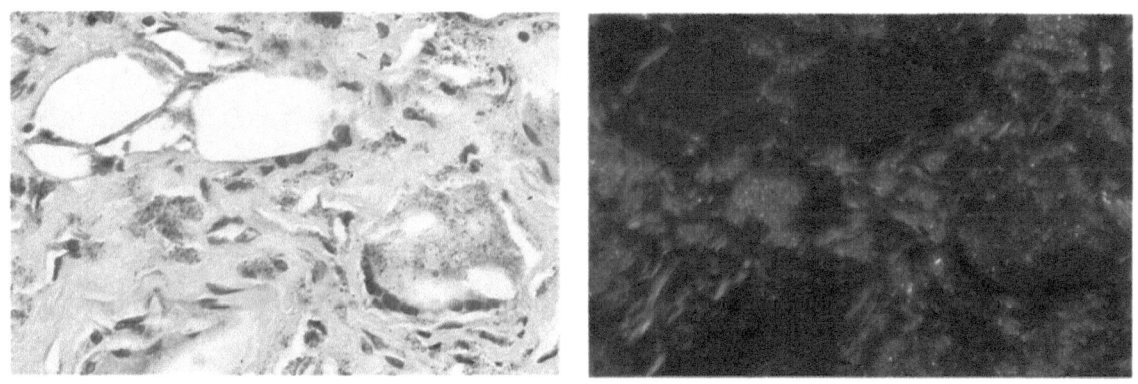

Fig. 10.5

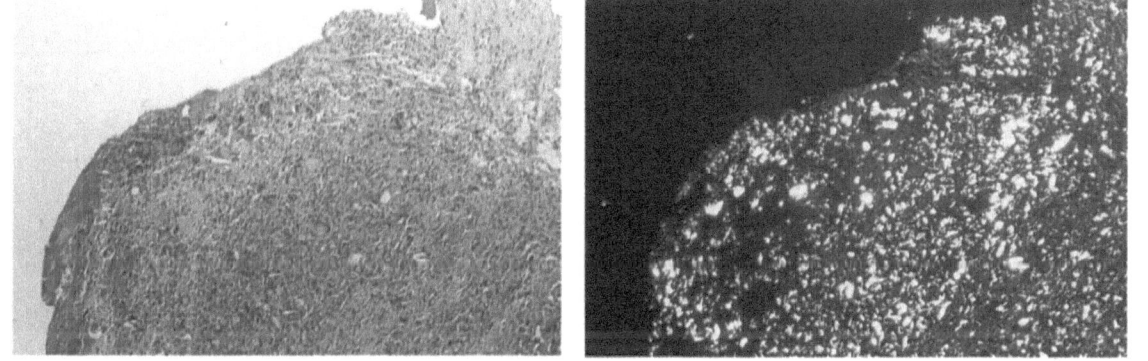

Fig. 10.7

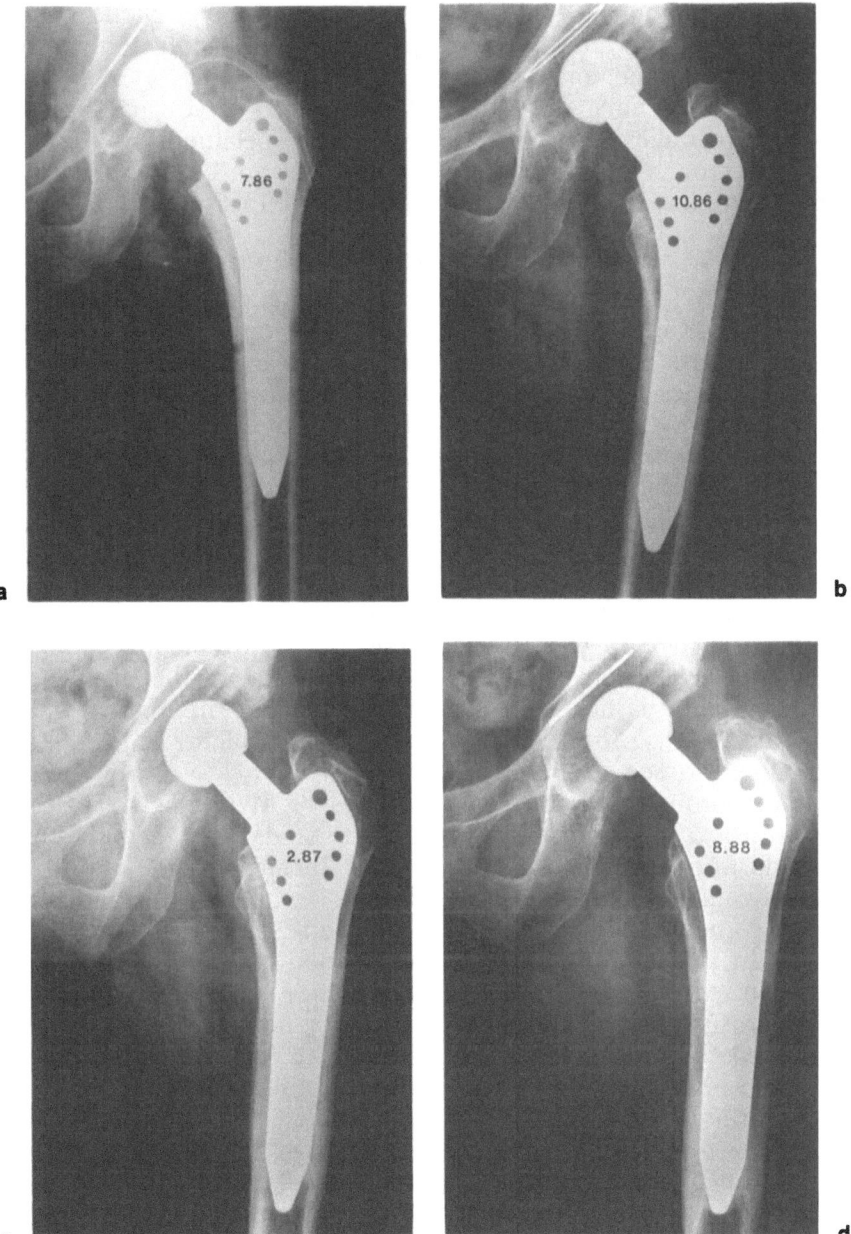

Fig. 10.8a–d. Osteolysis around the middle of non-cemented prosthetic shaft. Series of radiographs of a patient's left hip; follow-up over 2 years of function of a non-cemented total hip replacement with TiAlV forged alloy femoral stem, alumina ceramic ball head and a non-cemented all-polyethylene acetabular component. **a** Antero-posterior view a few days post-operatively. In this plane, the femoral component demonstrates excellent contact to cortical bone resulting in a stable press-fit. **b** Antero-posterior view 3 months post-operatively; adaptation of acetabular bone to load transfer by formation of sclerosis within the threads. **c** Antero-posterior view 7 months post-operatively; periosteal apposition of bone in the region of the tip of the stem, no sclerosis within the threads in caudal region of socket anchorage. **d** Antero-posterior view 2 years 1 month post-operatively; endosteal apposition of bone around the tip of the stem, loss of bone mineral around the middle of the prosthetic shaft more pronounced on the lateral side.

which is on the osseous surface and actually leads to resorption. When you look at your retrieved specimens – are you seeing macrophage resorption of bone, or is it all osteoclast? You and Roberts have published work suggesting that metallic debris was a greater stimulant of macrophage activity than polyethylene. Do you still feel that way or can you detect a difference?

Professor Howie: With regard to macrophages directly resorbing bone, I think the theory changes every 5 years or so, according to which studies are done and whether you are discussing real or pathological monocytes. One does see resorption of bone without many osteoclasts. Sometimes I wonder how one can have all this resorption and not see many osteoclasts. I suppose I am not yet ready to say that there is direct resorption by macrophages. We stained phosphates in giant cells which were well away from bone but did not really correlate with particles. We see also many particles and no bone resorption in the vicinity. So there is no good correlation. One can see lots of wear particles and a lot of fibrous tissue and very active resorption; so it is a three-dimensional problem as well. I do not think we said that metal particles were worse than polymer. It is almost impossible to compare the same size of particles of polymer in the same volumes in vitro or even in vivo. Metal particles cause more necrosis and fibrosis than polymer particles. That is my gut feeling from in vivo studies. I would probably rather see some fibrosis than possible continuous stimulation of macrophages to release cytokines and stimulate bone resorption. My gut feeling is to be a bit more concerned about polyethylene than metal particles.

Dr Isaac: Given that ceramic rather than polyethylene should produce better in vivo wear results, have you tried ceramic as opposed to polyethylene in any of the prostheses you have used, and managed to see whether there is a reduced amount of wear debris and any difference in size or shape?

Professor Willert: We are using the ceramic and polyethylene and have done so for more than 10 years now. We measured the migration of the head into the polyethylene cup and we found that the migration is about half in ceramic—polyethylene combination compared to metal—polyethylene combination.

Dr Isaac: Have you managed to look at any retrieved specimens to see if there is a reduced amount of wear debris and if there is any change in size or shape?

Professor Willert: No. In those cases with the ceramic—polyethylene combination, there was only a minor degree of polymer wear. This does not mean that there is a real difference in the action of polyethylene wear, but in the cases that we retrieved with a ceramic head, the problems with the bone cement particles within the tissue were more pronounced than in the polyethylene. But we have only very few cases.

Professor Slooff: The incidence of endosteal lysis is very low in my opinion. Is it possible that the cause of this disease could be failure in the cement mixing, so that the powder is really the problem?

Professor Willert: The cases that we were able to investigate came from different surgeons, including some of our own. We looked at the remnants of the bone cement and could not see any major disturbances in the remnants which were not worn or broken. I know that patients with massive osteolysis are quite rare, but we chose them in order to investigate the role of the agent which causes this osteolysis. Huddlestone did not do any morphological investigations but tried to find some statistical facts about the formation of these localised osteolyses. One of his conclusions was that obviously defects in the bone cement mantle and direct contact between the metal and the bone are significantly higher in his cases with osteolysis, so he sees some connection between these factors.

Mr Northmore-Ball: Professor Willert showed an example of massive osteolysis from a soft-top prosthesis, and the wear products in that case will have come, of course, from the outside of the polyethylene rubbing against cartilage and eventually against bone. Presumably the polyethylene particles in this case are very different in shape. I wonder whether they might have a completely different effect from the ones coming from an articulating surface.

Professor Willert: You are right. These particles are thicker, but we also have some cases with heavily worn polyethylene sockets from inside. In those cases, especially a long time after implantation and with polyethylene fatigue, the particles were large. It might have something to do with the amount of abrasion of this material, so that the more it abrades the bigger the particles become; the roughness increases by increasing abrasion.

Professor Howie: On the subject of particle size, we have felt that small ones were nasty. In the rat model of resorption we used large particles. I do not think it is a criticism of the model that one still saw the same phenomena. A multi-nuclear giant cell with particles is different from our earlier discussion about the cement—bone interface. Many of them are aggregates of macrophages which have retained their definition around these particles. I think they are fairly dangerous.

Professor Willert: If you mean the size of the particles, I should like to add that in polyester heads the particles were considerably smaller than the polyethylene particles, and we have the same reaction.

Mr Ling: I suggest that there may be another mechanism for the phenomenon that you describe. Where there is a defect in the cement mantle there is a potential communication between the stem and the cement, between the endosteal surface of the bone and the joint. There is a route through which polyethylene can get down into those lesions. I think we have proved also that the stem itself, particularly the loading and twisting of the stem, has an effect. If the stem surface is rough it will abrade the cement, and that is another source of acrylic particles which can get down between the stem and the cement and out through the hole in the cement mantle, and all the pressure changes in the joint are then communicated down to the endosteal surfaces. That may be one explanation for this localised resorption.

Chapter 11

Pathology of Cemented Low-Friction Arthroplasties in Autopsy Specimens

A.J. Malcolm

The late Sir John Charnley had the foresight to ask many of his patients to bequeath their hips to him. His intention was to follow up his initial study of the reaction of bone to cement during his retirement (Charnley 1970). Over the years he retrieved low-friction arthroplasties at autopsy storing them in formalin, but unfortunately he died before being able to fully examine them. I am grateful to have been given the opportunity to investigate these specimens. I will describe my findings and use them to discuss the possible tissue reactions in hip joint replacements that can lead to loosening.

The specimens received were already split down the middle with the cement mantle intact but with the metal prosthesis removed. All the specimens had slab X-rays to show the bone—cement interface and to define any areas of sclerosis or lysis. There were 78 specimens from 61 elderly patients, 42 from the right hip, 36 from the left hip, and all the arthroplasties had been performed by Charnley 8 to 22 years prior to the death of the patient. It is important to note that these patients all had good or excellent clinical results and died of causes unrelated to their hips.

The specimens were examined using slab radiographs, micro-radiographs, microscopy, scanning electron microscopy and occasionally transmission electron microscopy. Sections were taken down the femoral component to correspond with the seven radiological zones (Gruen et al. 1979). Sections were also prepared from the acetabular cup and the joint capsule. In order to convince sceptical orthopaedic surgeons that a bone—cement interface can occur, it was very important to retain and not dissolve out the cement in the specimens so that the bone—cement interface could be demonstrated. Routine processing in plastic or wax causes loss of the cement. A modified technique for embedding bone in plastic and cutting undecalcified sections was devised (Pallet et al. 1985). However, it was not possible to stain the cement so that it could be seen easily.

Femoral Component

Viable bone, together with fatty and haemopoietic marrow could be seen tight up against acrylic cement (Fig. 11.1). The cement was integrated with the medullary bone, and there were bone pegs jutting into the irregular outline of the cement. There had been remodelling of the medullary trabeculae with the bone struts arranged radially between the cement and the endosteum along the new stress lines placed upon the bone by the

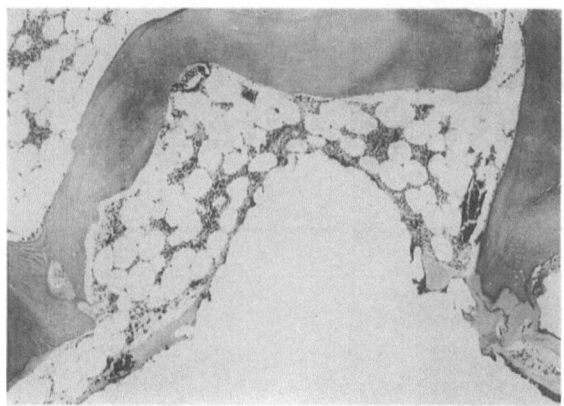

Fig. 11.1. Viable marrow and lamellar mineralised bone around a cement indentation with a bone peg (*right*) showing a layer of osteoid. (H & E, ×80)

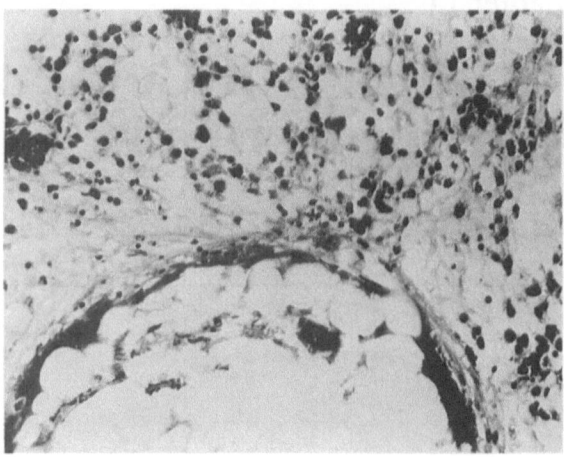

Fig. 11.4. Fatty and haemopoietic marrow (*above*) and bone cement (*below*). Scalloped giant cells between the cement spherules and marrow. (H & E, ×300)

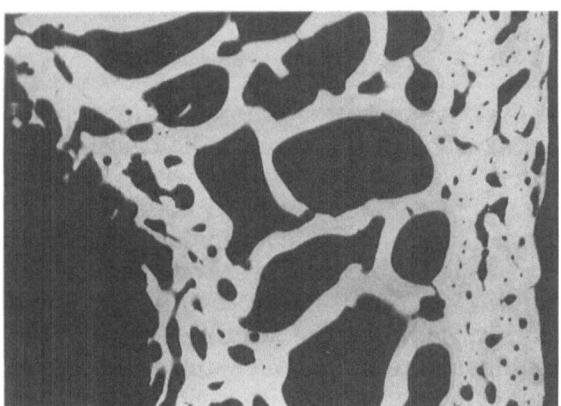

Fig. 11.2. Micro-radiograph of cortical bone (*right*) and medullary bone supporting a neo-cortex (*left*) with irregular bone pegs. (×20)

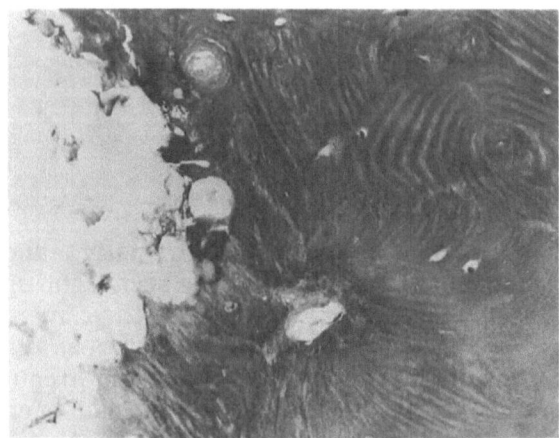

Fig. 11.3. Undecalcified histology showing slightly irregular lamellar mineralised bone, with osteocytes, integrated with cement (*left*) and showing cement spherule indentations. (Masson Trichrome, ×200)

implant (Fig. 11.2). It is noteworthy that normal haemopoietic marrow was present within a few microns of bone—cement. Haemopoietic tissue is very sensitive to any toxin and this finding strongly suggests that there is no cytotoxicity of intact cement. At a higher power, the irregular nature of the bone—cement interface could be seen. The bone had grown around the cement irregularities typified by the cement beads (Fig. 11.3). This is viable mineralised bone of a lamellar type integrating with the cement. There was no defect in bone mineralisation. In one or two areas, particularly where there was simply fatty and haemopoietic marrow, there were a few giant cells between the marrow and cement. These giant cells were irregular in shape reflecting the irregular cement outline (Fig. 11.4). There was no evidence of wear particles in any of these cells.

The arrangements of the bone and cement are important. Not only do they integrate but, as a three-dimensional reconstruction illustrated, the bone struts are arranged to support the mechanical forces exerted by the cement and the prosthesis. Wolff's law of transformation of bone suggested that if bone was appropriately stressed it would thicken and be laid down along those lines of stress (Wolff 1892). This is probably true in long bones. However, if the bone is under-stressed it is often resorbed causing the bone to become osteopenic. If bone is over-stressed, either resorption or fracture of the bone may occur.

If bone is appropriately stressed, there is a gradual remodelling of the bone as a reaction to that stress and the net result is thick, correctly orientated bone as shown in Fig. 11.2. The stress of an implant on bone is quite considerable. It has been estimated that there may be a stress on the femoral component of up to seven times the body weight when climbing stairs. Even this estimate may be low when additional rotational forces are exerted and shock loads occur, so that the bone may have to withstand considerably more peak stresses.

Very little medullary bone is present below the lesser trochanter in a normal femoral canal. There is an organised increase in this bone, however, after the implant of a cemented prosthesis. This new bone often forms a neo-cortex which is strutted on to the endosteum by multiple trabeculae as shown in Fig. 11.2. All this new bone has formed after implantation along the lines of stress as Wolff has dictated. This is more than osseo-integration; it is a good biological response because the bone laid down is appropriate to the stresses put upon it by the implant. So what is a good biological result? This study suggests that it is where viable mineralised bone integrates with the cement and is oriented along the lines of stress. This re-orientated bone supports the implant without fracturing or allowing excess movement. There is inevitably some movement between bone and cement on loading but when this is minimal there is no fibrosis at the interface between the cement and the bone.

Considering Charnley's crude technique and that these prostheses had been implanted and functioning for a long time, it is interesting that as many as 60 of the 78 specimens showed osseo-integration throughout almost the entire length of the prosthesis, with no evidence of a fibrous interface. In the gaps between the supporting bone pegs there was normal marrow in apposition with cement completing a biological "seal" around the cement mantle.

"Loosening" can be a clinical or radiological diagnosis. However, it may be "biological" when there is fibrous tissue between the cement and the bone which is too thin to be seen radiologically. There was a radiolucent area on slab X-ray in some sections taken down the femoral shaft which represented a

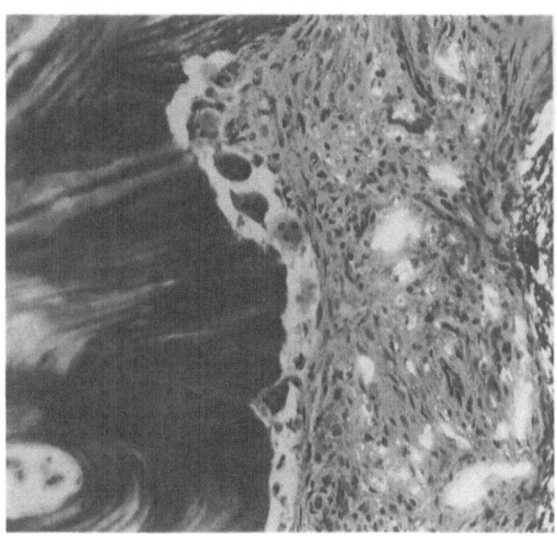

Fig. 11.5. Polarised light microscopy of an uncalcified section showing mineralised lamellar bone (*left*) and fibrosis with macrophages (*right*). High-density polyethylene wear particles within macrophages and osteoclasts resorbing bone. (Masson Trichrome, ×160)

thin fibrous membrane histologically; there was mineralised bone with a little osteoid, indicating that there was bone turnover. Instead of the bone integrating with cement, however, there was a significant fibrous membrane between them up to 2 mm in thickness. Macrophages were seen in some areas of the fibrous membrane and, by polarising microscopy, large numbers of high-density polyethylene wear particles could be demonstrated (Fig. 11.5).

Osteoclasts were often actively resorbing the bone immediately under this fibrous membrane (Fig. 11.5). In the 18 cases where there was a fibrous membrane between the bone and cement, this layer of fibrous tissue extended 6—15 cm from the proximal femur and wear particles were always present.

There was significant osteoporosis in some of the specimens, not unnaturally considering the patients' ages, and the supporting cross-struts of the bone trabeculae had become rather thin. There were fractures of these trabeculae in some areas (Fig. 11.6). These fractures were healing, not by reactive new bone being formed but by the formation of cartilaginous tissue. This is identical to the features seen in non-union and suggests that continuing movement was occurring at these

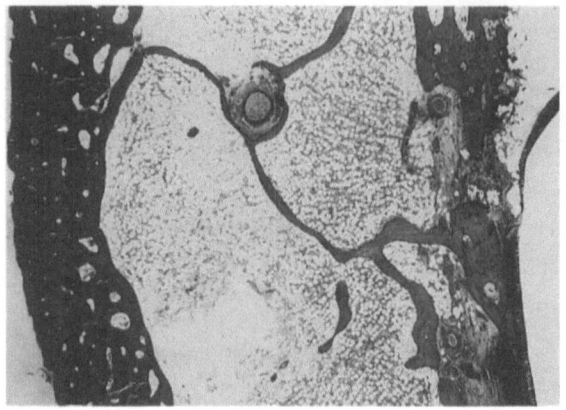

Fig. 11.6. Undecalcified section of cortical bone (*left*), marrow, a fractured medullary trabeculum and neo-cortical bone (*right*). There is a fibrous membrane between bone and cement (*extreme right*). (Masson Trichrome, ×20)

fracture sites, thus preventing normal fracture repair. A fibrous membrane at the bone—cement interface is seen in Fig. 11.6. When stressed, excessive movement would occur between the bone and the cement because of the fractured trabeculae, causing a fibrous membrane to form between bone and cement resulting in loss of osseo-integration. Eight of the 18 cases showing a fibrous membrane at the bone—cement interface had microfractures of supporting trabeculae.

Acetabular Component

Two main types of high-density polyethylene cups were inserted. One was cemented and the other was a metal-backed cup. I will describe the tissue reaction to cemented cups since the metal-backed cups had an identical fibrous membrane between the cup and bone as was found in the cemented cups. Every single cemented cup showed a fibrous membrane between the bone and cement and none showed the osseo-integration found in the femoral component. Microscopy showed wear particles within a fibrous membrane, numerous macrophages and many osteoclasts actively removing the bone at the interface with an appearance identical to that shown in Fig. 11.5. This was a consistent finding in every acetabular cup.

Joint Capsule

The fibrous tissue of the capsule showed metal wear particles and some necrosis. There were numerous shards of high-density polyethylene, some of which were too large to be ingested by macrophages and were surrounded by multinucleated cells containing some quite small high-density polyethylene wear particles. These particles arise from the inevitable wear between the head and the cup. There are varying wear rates, producing enormous amounts of wear particles. The effect of these wear particles will depend upon their size and their chemical nature. Some are cytotoxic. Small particles can be ingested by macrophages and transported to different sites while larger particles tend to remain in the capsule. These findings have been widely reported (Eftekhar et al. 1983, Revell 1982).

Conclusions

Osteoclastic resorption of bone can be stimulated by a variety of factors including infection, inflammation and excess movement. Inflammation can be due to either corrosion or wear particles. It is known that an activated macrophage associated with ingestion of wear particles can release very many substances, among them prostaglandins, interleukin-1 and tumour necrosis factors, all of which actively stimulate osteoclastic activity (Thomson et al. 1986). It is reasonable to suggest that wear particles ingested by macrophages may cause these macrophages to release osteoclast-activating factors and, therefore, initiate bone resorption.

All biological implants are going to produce wear particles which will accumulate in the capsule. Some will be removed, others will not. When the wear particles build up in close apposition to bone, the macrophages may release substances which can activate osteoclasts to remove that bone. When osteoclasts remove bone from the bone—cement interface, this allows the ingrowth of granulation tissue and fibrous tissue at the interface. Wear particles may enter into this fibrous tissue either within the flux of fluid that occurs between this fibrous layer and the joint fluid when cyclically loaded, or be transported within migrating macrophages. When wear particles appear in this fibrous membrane between the bone and cement they attract macrophages which can excite osteoclasts to remove more bone. This mechanism may progress around the acetabular cup and down the femoral component. The inevitable result is that insufficient bone is present at the interface for the prosthesis to remain secure. Movement will start to take place between cement and bone and an inevitable cycle of loosening may occur. Wear particles, therefore, may initiate bone loss or may exacerbate bone loss caused by other factors.

However, rather than blaming wear particles exclusively, there are a variety of other mechanisms that may be implicated. Trabecular microfractures in the supporting bone may, in certain situations, cause loosening. If a patient suddenly stresses an implant sufficiently so that multiple adjacent supporting trabeculae all fracture, and this stress is continued, the prosthesis may no longer be held rigid resulting in movement. If there is excessive movement, bone repair and bone ingrowth cannot occur and therefore a fibrous interface may form. This fibrous layer may allow migration of wear particles up and down the interface as already described. These microfractures are more likely to occur in middle-aged and elderly patients in whom it is known that the medullary canal enlarges (Ruff and Hayes 1982). This enlargement of the canal results in the prosthesis being directly supported by more medullary bone and less endosteal bone. This occurs at a time of increasing osteopenia thus further reducing the strength of the supporting bone. Equally, a fracture of the cement mantle may also allow movement to occur between bone and cement which may result in the same net loss of bone and influx of wear particles.

However, it is remarkable that 60 of the 78 femoral components had an intact bone—cement interface throughout. There were no fibrous membranes or foreign body giant cells, and the prostheses had functioned very well for up to 22 years. All the acetabular cups had a fibrous membrane between the bone and the cement yet all had apparently functioned well.

References and Further Reading

Charnley J (1970) The reaction of bone to self-curing acrylic cement. A long-term histological study in man. J Bone Joint Surg (Br) 52:340—347

Eftekhar NS, Doty SB, Parisien MV (1983) Prosthetic synovitis. Hip 13:169—183

Gruen TA, McNeice GM, Amstutz HC (1979) "Models of failure" of cemented stem type femoral components. A radiological analysis of loosening. Clin Orthop 141:17—27

Pallet CD, Mawhinney WHB, Malcolm J (1985) Plastic processing of cemented hip joint replacement specimens. J Clin Pathol 39:339—342

Revell PA (1982) Tissue reactions to joint prostheses and the products of wear and corrosion. In: Berry CL (ed) Bone and joint disease. Springer-Verlag, Berlin Heidelberg New York, pp 73—101

Ruff CB, Hayes WC (1982) Subperiosteal expansion and cortical remodelling of the lumen femur and tibia with ageing. Science 217:945—951

Thomson BM, Saklatvala J, Chambers TS (1986) Osteoclast-mediated interleukin-1 stimulation of bone resorption by rat osteoclasts. J Exp Med 164:107—112

Wolff J (1892) Das Gesetz er Transformation knocken. A Hirschwald, Berlin

Discussion

The Chairman - **Dr Linder**

The Panel - **Dr Malcolm**
        - **Dr Draenert**

Mr Ling: In how many of these cases did you find fractured cement?

Dr Malcolm: I did not find fractured cement that I could see grossly just by looking at it. At the time, however, I was somewhat tunnel-visioned into cement bone. I have not performed any technique such as ink-staining or scanning electron microscopy to see how many of those cement mantles have any fractures.

Professor Fitzgerald: Harris thinks the breakdown is occurring at the cement—prosthetic interface. Contrary to Ling he thinks that a rough or pre-coated surface is advantageous because it maintains the prosthetic interface which he thinks is more critical.

Mr Ling: I know his views, of course — time will tell!

Professor Slooff: Do the regions where you found the cartilage cells correspond with the absence of vessels?

Dr Malcolm: No, there was no evidence of avascularity. The bone either side of the fracture site still had osteocytes and there was a foamy macrophage reaction to the disrupted fat. I took both of those features to indicate that the blood supply was still intact.

Mr Elson: Everyone is accepting readily that polyethylene debris is a prime producer of loosening. Perhaps Dr Draenert would comment.

Dr Draenert: We found osteolytic lesions near the newly formed capsule in our 39 specimens. One level below the end of the osteolytic lesion the cement sheath was completely intact, however. We do not believe, therefore, that loosening is caused by wear particles produced by polyethylene.

Dr Grobbelaar: We have been looking at this same problem for many years. There is a definite pattern and a difference between the polyethylene and the acrylic reactions. The acrylic reaction takes place mainly around stems which then tend to become loose quite rapidly. The cystic reaction that Dr Draenert has just described will be there initially without any looseness. These patients usually have very little or no pain at that stage, but the cystic reaction eventually becomes so widespread that the implant will become loose. A very large cavitation then surrounds the implant.

I think that polyethylene is definitely the only basic cause of loosening. Later the acrylic debris will also be there because of fretting of the acrylic against the bone.

Mr Miles: Did Dr Malcolm see any regional differences in those trabecular struts that he noted which may be indicative of the mechanical environment locally — for example, on the proximal medial portion or the distal lateral?

Dr Malcolm: We have not had the time or the opportunity to take multiple sections all the way down and do a three-dimensional reconstruction. Nevertheless, just looking at the sections we have, the pattern of bone stress is much as one would predict. The presence of thick versus thin trabeculae in the same patient is indicative of the stresses and strains; the superior medial and distal lateral portions contain most bone, which is perhaps all that one might expect.

Dr Linder: Do you have a feeling that different types of interface such as direct bone—implant contact or fibrous interface provide different resistance to the spread of wear?

Dr Malcolm: Clearly, I am only looking at one snap moment in time. One interpretation is, however, that where there is a bone—cement interface there is fatty and haemopoietic marrow between the bone pegs. This effectively is like a biological seal for preventing wear particles from travelling down. Where you get a fibrous interface with fluid flux and the ability of cells to move within it, you start to run a risk of further transport of wear particles.

Part II

Bone Cement

Chapter 12

Time-Dependent Properties of Polymethylmethacrylate Bone Cement

A.J.C. Lee, R.D. Perkins and R.S.M. Ling

Introduction

Although the static and fatigue properties of PMMA bone cement have been extensively studied (Saha and Pal 1984) relatively little attention has been given so far to the time-dependent properties of the material, especially under environmental conditions similar to those to which it is exposed in the context of total hip arthroplasty "in-vivo".

Creep is most simply defined as time- and temperature-dependent deformation under constant load. Stress relaxation (diminution of stress level at constant strain), which is also time and temperature dependent, is a characteristic associated with materials that creep and can be regarded as the "inverse" of creep.

Most materials, and particularly metals, exhibit a stress/strain characteristic at body temperature for a given load that is independent of time. The stress in a metal is directly proportional to the strain and does not vary with the amount of time the load has been applied. Some materials, and especially polymeric materials, exhibit a stress/strain characteristic that is not independent of time at body temperature and these show a slow, continuous deformation under constant load (they creep at body temperature). This paper describes the creep behaviour of self-curing polymethylmethacrylate bone cement at room temperature and body temperature under a number of loading conditions.

The phenomenon of creep in polymers needs some explanation. A temperature known as the glass transition temperature (Tg) is close to room temperature for many polymers. Well below Tg, the polymer is a glass and a brittle elastic solid. Well above Tg, a thermoplastic polymer such as PMMA is a viscous liquid. Body temperature (310°K) is very close to the glass transition temperature of PMMA (373°K) so PMMA at body temperature behaves in a manner that is neither that of a simple elastic solid nor that of a true viscous liquid. It behaves as a visco-elastic solid; the behaviour of such a solid can be represented by a coupled spring and dashpot (Fig. 12.1). The elastic behaviour is represented by the spring and the viscous behaviour by the dashpot. Applying a load causes a small elastic deformation, followed by continuing deformation (creep) at an ever decreasing rate, because the spring is taking up the tension. Releasing the load causes slow reverse creep

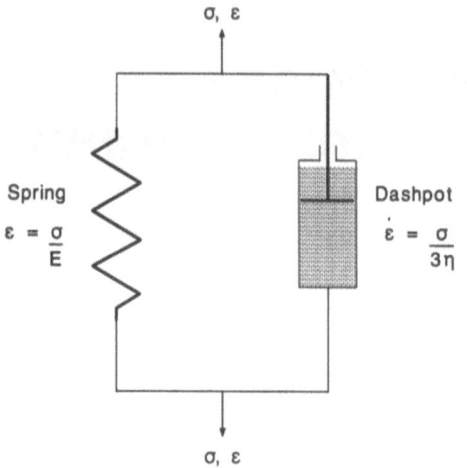

Fig. 12.1. Spring and dashpot model to describe creep in polymers.

because the extended spring regains its original length against the resistance of the dashpot.

Previous Work

Results of creep investigations on commercially available bone cements were first published by Treharne and Brown in 1975. Compression creep and recovery tests were performed at room temperature (22.8—26.1°C). They established for Surgical Simplex Plain and Radio-opaque cements that the greater the cement density the less the creep; radio-opaque cement creeps slightly less than radiolucent cement; cements with larger particle sizes or additions of MMA-styrene copolymer creep less; aqueous solutions plasticise cements to give more creep; and creep strength is lower when residual monomer is higher. Creep and stress relaxation tests on CMW, Surgical Simplex and Palacos bone cements have been described (Holm 1980). Testing was carried out at body temperature, 38°C, using specimens of acrylic cement moulded across the open end of a "U"-shaped spring that was designed to be deformed, thus applying a load to the cement that decreased as the spring regained its original shape.

The creep phenomenon was described as a stretching and re-aligning of molecules of the polymeric bone cement which is greatly facilitated by the presence of water acting as an internal "lubricant". The degree of polymerisation and the extent of porosity of the cement were found to be important. The results of the creep and stress relaxation tests showed that CMW and Simplex behave in a similar way, while Palacos behaved in a significantly different manner. Both CMW and Simplex stress relaxed to lose between 20% and 30% of initial stress after 500 hours; Palacos had lost 60% of its initial stress by this time. Simplex and CMW relaxed to a minimal stress of 0.1 N/mm^2 in 1 to 2 years; Palacos relaxed to this level in about 1000 hours. Long-term tests on CMW showed, as expected, that the rate of stress relaxation becomes slower as the stress gets smaller. Compression creep tests at 37°C were carried out on five types of cement (Zimmer Regular, LVC, carbon-reinforced, Omniplastic and Surgical Simplex) (Chwirut 1984). Stress levels were set at "worst clinical" and tests lasted 1000 hours. Differences were observed between cements; Zimmer carbon-reinforced cement crept least, followed by Simplex and Zimmer LVC, with Zimmer regular and Omniplastic creeping twice as much as Simplex and LVC. Various authors have investigated the effect of adding fibre reinforcement or hydroxyapatite to the cement using room temperature compression creep and stress relaxation tests (Pal and Saha 1982) and 37°C creep bending tests (Castaldini and Cavallini 1986). It was found that the addition of fibre reinforcement reduced creep significantly; conversely, voids in the cement increased creep significantly. The addition of hydroxyapatite modified the pattern by reducing voids and so increased creep resistance. An experimental investigation into the gap between cement and stem in a total hip replacement was carried out (Ebramzadeh et al. 1985).

Charnley-type implants were cemented into polyacetal tubes with strain gauges embedded in the acrylic cement. Testing was at room temperature in air. Stress levels high enough to cause creep of cement were observed in the proximal medial region. After cyclic load testing designed to give a stress of about 10 MPa in this region, it was found that stress relaxation in the cement had caused the stress level to reduce to 27% of the original level after 4 million load cycles. Gaps were observed between cement and stem similar to those that have been seen after standard fatigue tests of stems cemented at varying heights into constrained cement. No cracks were observed in the cement.

A new bone cement was tested consisting of polymethylmethacrylate powder with 1.5% w/w benzoyl peroxide and barium sulphate added; the monomer was *n*-butyl methacrylate with the addition of 2.5% v/v *N,N*-dimethyl *p*-toluidine (Weightman et al. 1987). This cement was shown to creep to a very much greater extent than conventional PMMA bone cement when tested in a simplistic stem model and when tested with commercially available stems in cadaveric femora. These authors made the important statement that a cement less creep resistant than PMMA, when used with femoral components of a suitable design, could improve the clinical situation.

A recent paper presented at the Third World Biomaterials Congress in Japan confirms that PMMA bone cement behaves as a linear visco-elastic polymer (McKellop et al. 1988). At the same Congress, investigations into the properties of a rubber-modified cement were presented (Murakami et al. 1988). The creep of this cement was 10 times that of CMW cement tested at the same time; this was said to be a disadvantage of the cement. McKellop and co-workers also showed that PBMA cement tested dry and at room temperature exhibited 14 times the continuous flow deformation of PMMA. They concluded that the ideal combination of elastic modulus and viscous flow characteristics may lie between those of the two types of cement tested (McKellop et al. 1989).

Materials and Methods

The authors have tested the effect of a number of variables on the creep of PMMA cement using four-point bending tests (Fig. 12.2) at the relatively high stress level of 32 N/mm². The variables assessed were: mixing by hand or using a vacuum mixer, the effect of storage and testing temperature (20°C or 37°C) and the effect of age (by testing specimens that had been stored for 1 week, 3

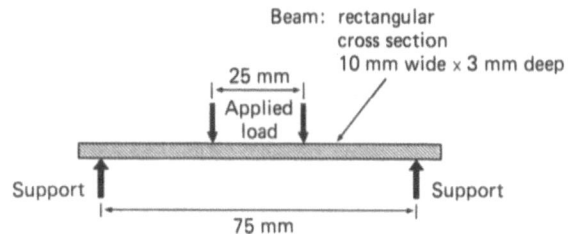

Fig. 12.2. Four-point bending.

weeks or 6 weeks, either dry or in a saline solution). A further series of tests at stress levels of 15 N/mm² and 5 N/mm² were carried out to assess the effect of stress level.

In order to mimic the medullary fat environment in which bone cement normally operates, a series of tests were performed in which the specimens were stored and tested in intralipid.

The above described tests were all performed with a constant static load applied to the specimen. "In-vivo", the load applied to cement via the implant is neither constant nor static, but is variable and repeated in a cyclic manner (Berme and Paul 1971). Consequently, a series of tests was performed in which the applied load was cycled in a sinusoidal manner between 0 and 40 N.

In order to extend the testing to a situation that more closely resembled that of the stem in a femur, a series of tests was performed in which a conical shell of cement, constrained around the outside, was loaded by a central taper in compression.

Results

Four-Point Bending Tests

The classical pattern of creep in normal specimens is shown in Fig. 12.3. When the initial load is put on the specimen there is an immediate elastic deformation. The specimen then extends fairly rapidly under constant stress (primary creep), easing into a steady deflection rate (secondary creep) followed finally by an increased creep rate (tertiary creep) which leads to specimen failure. Figure 12.4 shows the results of a set of four-point bending tests of specimens stored and tested in isotonic saline at 37°C at various times after specimen manufacture.

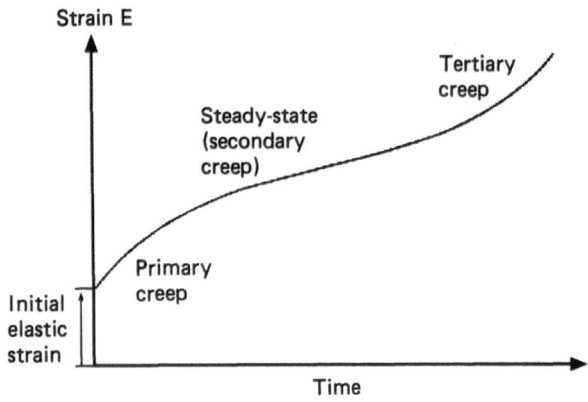

Fig. 12.3. Typical creep curve.

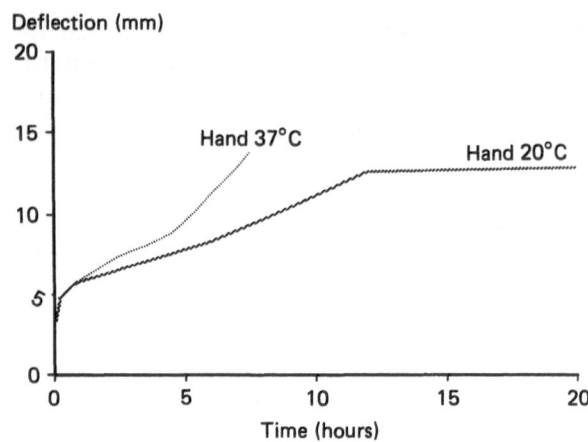

Fig. 12.5. Creep of PMMA: four-point bending in saline: 37°C vs 20°C.

It is clear that creep occurs under high stress conditions. When the stress levels were reduced to nearer the levels found in total joint replacements (15 N/mm² and 5 N/mm²), creep was still observed of the same general form but at a much reduced rate. Even with physiological load conditions, therefore, creep of bone cement does occur.

In a further experiment the bone cement was mixed and the specimens stored dry at 37°C. Storing the specimens in saline tends to plasticise the bone cement so the dry-stored specimens do not creep at the same rate: dry specimens creep more slowly than wet specimens.

The ambient temperature probably has the most significant effect on the creep rate of PMMA bone cement. Testing at room temper-

ature (20°C) shows that creep rate is much slower than at body temperature (37°C) (Fig. 12.5).

In a series of tests designed to emulate the body's environment more closely, the specimens were stored and tested in Intralipid at 37°C, with extremely interesting results. Tests were carried out after various storage intervals. As expected, creep at 7 days was slower than at 2 days. However, creep at 21 days and 42 days showed a reverse trend, with creep rates increasing (Fig. 12.6). This indicated that bone cement stored and tested in a solution which contains fat may be plasticised to a considerable extent over a relatively short period of time.

The preliminary conclusion from these tests is cement will creep at realistic stress levels

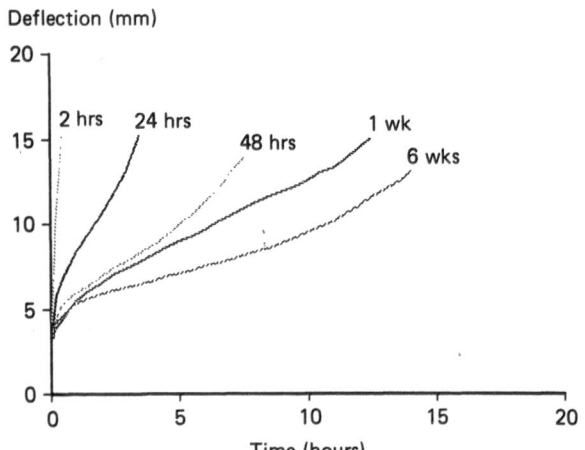

Fig. 12.4. Creep of PMMA: four-point bending in saline at 37°C.

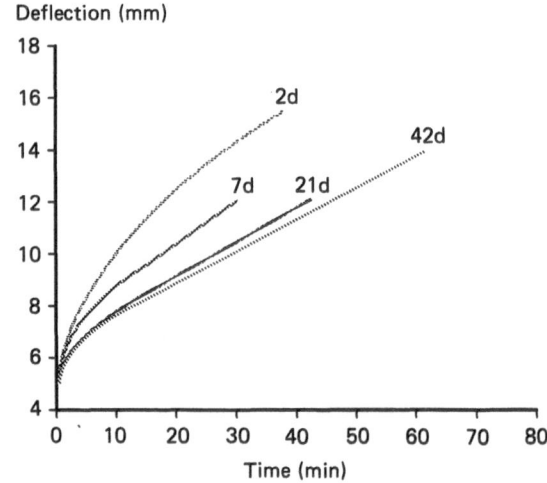

Fig. 12.6. Creep of PMMA: four-point bending in intralipid at 37°C.

and that it is very sensitive to the ambient temperature and environment. When bone cement is implanted in a patient, it is likely to creep much more readily than most investigators previously believed.

Tests were made to determine whether a period of creep affected the final strength of the material. Twelve specimens were made; all were mixed in an identical way, stored at 37°C in saline and then tested to destruction in four-point bending 48 hours after mixing. Six specimens were allowed to creep for 6 hours before testing to destruction. The other specimens were not allowed to creep before testing. The failure loads were measured and found to be 69.8 N with no creep and 67.5 N with creep, a difference that was not statistically significant. This establishes that, although bone cement will creep, its final rupture strength is not significantly affected.

In vivo cement is loaded in a repeated cyclic manner, from heel strike to toe off. In order to determine the effect of cyclic, as opposed to static, loading, specimens were placed in a Dartec testing machine with the load cycling between 0 and 40 newtons. An automatic print out of deflection against time was obtained that showed cement would creep under cyclic conditions (Fig. 12.7). The tests showed that the age of the specimen has an effect on creep rate under cyclic as well as static loading conditions.

Taper Constraint Tests

The investigation extended to a model that more nearly represented the stress conditions that bone cement might encounter inside a patient. The test shown diagrammatically in Fig. 12.8 was set up: a conical metal constraint enclosed a conical shell of bone cement that had a polished conical metal taper in the middle. A 10 kN constant load was put on the taper. The experimental apparatus was maintained at an ambient temperature of 37°C. After testing under load for 48 hours, the bone cement was observed to have extruded from the end of the metal constraint. The specimen was removed from the apparatus and a small longitudinal crack was found in the extruded portion; the constrained portion was not cracked anywhere. This implies that the stress situation inside a constrained

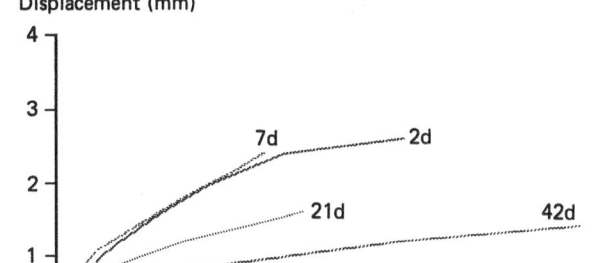

Fig. 12.7. Creep of PMMA: four-point bending cyclic loading at 37°C.

tapered mantle of cement will enable the cement to survive high loads without failure. Where cement is not constrained (in the extruded portion) the stress conditions may be such that the tensile hoop stress is able to cause cracking.

A second set of taper constraint tests was carried out in which the outer metal constraint had a series of grooves cut into it (Fig. 12.9). A plain, smooth, conical shell of bone cement was placed in the constraint, and the central smooth polished metal taper forced down the middle as before. As the taper crept down the centre under constant load, the bone cement crept radially outwards into the grooves, with extrusion at the bottom as before. These tests showed quite clearly that

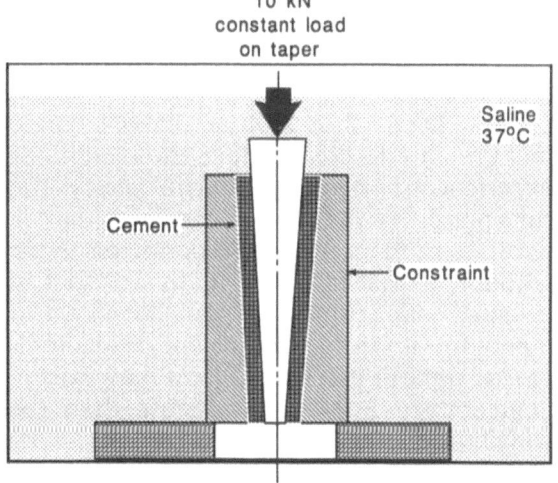

Fig. 12.8. Taper constraint tests: smooth constraint.

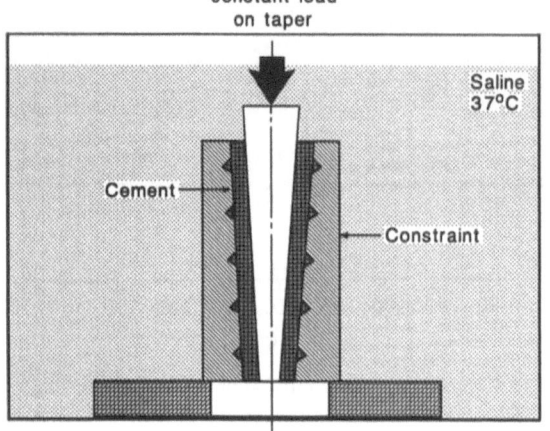

10 kN
constant load
on taper

Saline
37°C

Cement

Constraint

Fig. 12.9. Taper constraint tests: grooved constraint.

bone cement is able to move, by creep, in a way that can change the shape of the cement mantle and may change the contact conditions between cement and constraining bone.

Conclusions

The creep properties of cement have been measured in a number of laboratory tests which largely reflect a mixture of tensile and compressive creep, so that the various creep constants cannot be derived in a straightforward manner. Further tests are now under way in the authors' laboratory to clarify the nature of cement creep under tension and compression alone, and to derive the various creep constants.

It has been shown that PMMA bone cement is able to creep under the conditions it experiences in a total joint arthroplasty. The results presented in this paper indicate that all tests of cement need to be conducted in an environment that is representative of that in the living patient. In particular, the effect of temperature is critical, and recent results indicate that cement in a loaded joint may exist at a temperature 2°C or 3°C higher than the normally assumed temperature of 37°C (Humphreys et al. 1989). This fact may be significant in the design of standard tests used to evaluate implant performance.

The effect of creep and its associated property of stress relaxation means that, in addition to acting as a load spreader, shock absorber and load decoupler, the material may act so as to change the nature of the stress conditions at the bone—cement interface. With appropriate implant design this property can be used to the advantage of the whole system.

Acknowledgement. Dr Lee, who presented this paper, is indebted to his co-authors for their collaborative work. Mr R.D. Perkins was supported by the Freddy Durbin Charitable Foundation to whom thanks are given.

References and Further Reading

Berme N, Paul JP (1971) Load actions transmitted by implants. J Biomed Eng 1:268—272

Castaldini A, Cavallini A (1986) Creep behaviour of composite bone cement. In: Christel P, Meunier A, Lee AJC (eds) Biological and biomedical performance of biomaterials. Elsevier, Amsterdam, pp 525—530

Chwirut DJ (1984) Long term compressive creep deformation and damage in acrylic bone cements. J Biomed Mater Res 18:25—37

Ebramzadeh E, Mina-Araghi M, Clarke IC, Ashford R (1985) Loosening of well-cemented total hip femoral prostheses due to creep of the cement. In: Fraker AC, Griffin CD (eds) Corrosion and degradation of implant materials: Second symposium, ASTM STP 859. ASTM, Philadelphia, pp 373—399

Holm NJ (1980) The relaxation of some acrylic bone cements. Acta Orthop Scand 51:727—731

Humphreys PK, Orr JF, Bahrani AS (1989) An investigation into the effect of cyclic loading and frequency on the temperature of PMMA bone cement in hip prostheses. J Eng Med 203:167—170

McKellop H, Narayan S, Ebramzadeh E, Sarmiento A (1988) Visco-elastic creep properties of PMMA surgical cement. Trans Third World Biomaterials Congress, Kyoto, Japan, p 328

McKellop H, Narayan S, Lu B, Sarmiento A (1989) Visco-elastic properties of high and low modulus acrylic cement. Trans ORS 1989

Murakami A, Behiri J, Bonfield W (1988) Rubber modified bone cement. Trans Third World Biomaterials Congress, Kyoto, Japan, p 334

Pal S, Saha S (1982) Stress relaxation and creep behaviour of normal and carbon fibre reinforced acrylic bone cement. Biomaterials 3:93—96

Saha S, Pal S (1984) Mechanical properties of bone cement: A review. J Biomed Mater Res 18:435—462

Treharne RW, Brown N (1975) Factors influencing the creep behaviour of polymethylmethacrylate cements. J Biomed Mater Res Symp 6:81—88

Weightman B, Freeman MAR, Revell PA, Braden M, Albrektsson BEJ, Carlsson LV (1987) The mechanical properties of cement and loosening of the femoral component of hip replacements. J Bone Joint Surg (Br) 69:558—564

Chapter 13

The Scientific Basis of Vacuum Application of Bone Cement

K. Draenert

The success of implant fixation by bone cement is based mainly on enormous surface enlargement, reduced loading of bone and stiffening of the bone (Charnley 1970; Draenert 1986a). The stiffening of the cancellous bone framework is probably the most important contribution of bone cement to long-lasting results. The histological findings are perfectly explained by the deformation pattern of bone. The structure of the cancellous bone appeared unchanged after 8 years stiffened by bone cement. Study of areas of direct contact between bone and metal where there was no cement sheath revealed circumscript bone necrosis, microfractures or a fibrous tissue layer between the implant and the bony bed (Draenert 1988). The underlying bone was preserved by a thin cement layer in between the metal and the compact bone.

The effect of the cancellous bone stiffening can be demonstrated by a simple model. An empty egg will break if loaded by 2 kg, but no deformation is visible when the egg is filled with bone cement. This emphasises the importance of cementing technique (Lee and Ling 1981). Other study groups have tried to improve the physical properties of bone cement. A biological approach was taken by de Wijn (1976) who used gel to stimulate bone ingrowth; later calcium phosphates were used (Rijke and Rieger 1977). The surface properties of bone cement and bony ingrowth can be improved very elegantly using tricalcium phosphate and hydroxyapatite bone cements (Draenert 1986b). Different methods were used to improve the strength of the material. Centrifugation of low-viscosity PMMA cements led to a separation of their components, especially the radiological contrast medium (Dingeldein and Wahlig 1987). The introduction of fibre reinforcement showed the lattice phenomenon with the cross-anchoring pegs free of fibres (Pilliar et al. 1976). The most elegant method involved mixing PMMA bone cements under vacuum (Lidgren et al. 1984). This led to a 40% increase in fatigue strength.

The vacuum-packaged powder system produces a poor cement (Tepic and Perren 1985). All the polymer spheres are in contact with each other and there is no space for filler particles or even gentamicin.

The scanning electron microscope has shown that the polymer spheres of the powder component are not regularly moistened and embedded during polymerisation. Therefore it was concluded that the pre-pressurisation of PMMA bone cements was absolutely necessary (Draenert 1984).

Thrombo-embolism is the major complication and the leading cause of death during total hip replacement (Harris et al. 1977). The continuous monitoring of a patient during a

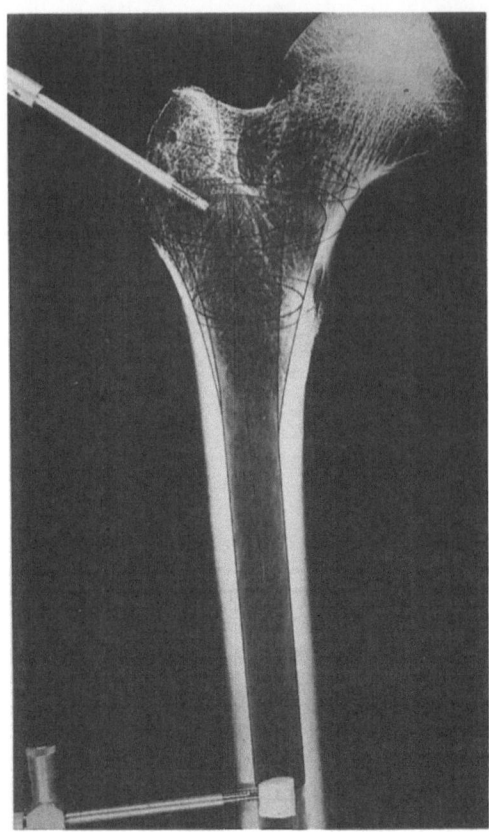

The different steps were carefully investigated in the laboratory and demonstrated in more than 33 hip courses using cadaveric specimens (Draenert 1988).

Conclusion

If vacuum methods are to be used in the mixing and insertion of cement, it is important that a precise sequence should be followed in the procedure. This involves the storage of normal viscosity bone cements in a refrigerator box (1—4°C); mixing in two steps under high vacuum (100 mbar), 15 seconds turbulently (4 times/sec), 15 seconds slowly (2—3 times/sec), 15 seconds evacuating without any movement; the extrusion of bone cement without hard contact; pre-pressurising bone cement (3—5 bar) for at least 1 minute; and the insertion of bone cement using the same

Fig. 13.1. The most elegant method of keeping the floor of the bone dry and of filling the cavity up with cement in an artefact-free manner without endangering the life of the patient by inducing a temporary increase in the intramedullary pressure is the process of vacuum filling, which takes only 3 seconds.

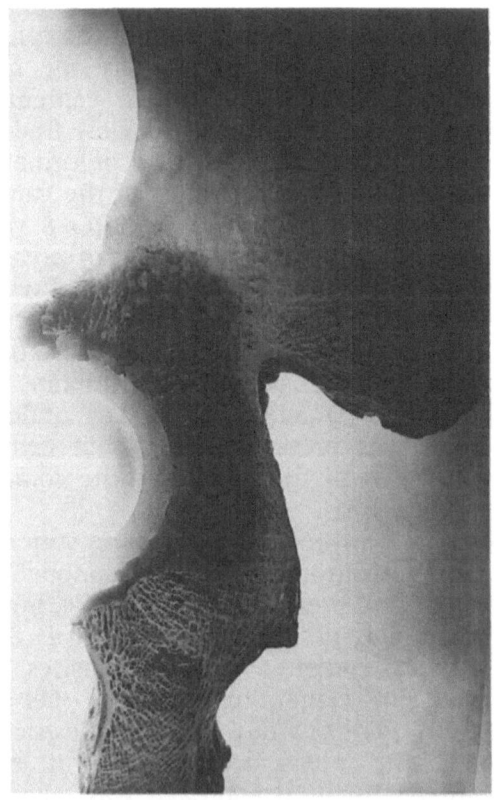

total hip replacement operation, including evaluation of changes in the right atrium and ventricle by means of trans-oesophageal two-dimensional echocardiography, elucidated the pathophysiology of the increase in intramedullary pressure and lung embolism (Ulrich et al. 1985).

A completely new technique was developed applying an effective vacuum during mixing and application of PMMA bone cements to improve the properties of bone cement and the cementing technique in order to avoid endangering the patient's life.

The instruments comprise a sterilisable vacuum pump with a very small deadspace volume for mixing and vacuum insertion of bone cement. A sterilisable disposable set was designed for simple handling, the results of which are reproducible.

Fig. 13.2. In the pelvis a special advantage is given by controlling the penetration depth of bone cement via the positioning of the tip of the cannulated screw.

vacuum pump via a cannulated screw and a cortico-cancellous plug as a filter system in the femur (Fig. 13.1). In the acetabulum, multiple holes made with diamond cutting tubes enhance the effect of the vacuum on the intrusion of the cement mass into bone.

No thrombo-embolism has so far been detected (Ulrich, unpublished data). The advantage of the vacuum technique is the rapid insertion of cement without any increase of the intramedullary pressure, keeping the floor of the bone dry and filling the cavity without artefacts. In the pelvis, positioning of the tip of the cannulated screw gives a special advantage for it controls the depth of penetration of bone cement into bone (Fig. 13.2).

References and Further Reading

Charnley J (1970) The reaction of bone to self-curing acrylic cement: a long term histological study in man. J Bone Joint Surg (Br) 52:340—353

de Wijn JR (1976) Polymethylmethacrylate aqueous phase blends: in situ curing materials. J Biomed Mater Res 10:625—635

Dingeldein E, Wahlig H (1987) The effect of centrifugation on radiopaque materials and antibiotics admixed to bone cements. In: Draenert K, Rutt A (eds) Beitrage zur Implantatverankerung. Histomorph Bewegungsapp 3:105—110

Draenert K (1984) Vorrichtung und Verfahren zum Mischen und Applizieren von Knochenzement. Offenlegungsschrift DE 3425 566 A1 vom 11.07.1984

Draenert K (1986a) Histomorphologische Befunde zur gedaempften und ungedaempften Krafteinleitung in das knocherne Lager. Vereinigung Nordwestdeutscher Orthopaeden. 36 Jahrestagung 15—18 Juni, Hannover

Draenert K (1986b) Histomorphological observations on experiments to improve the bone-to-cement contact. Nicholas Audry Award Paper presented at the Thirty-eighth Annual Meeting of the Association of Bone and Joint Surgeons, held in Vancouver, March 27—31. To be published in Clin Orthop, Lippincott, Philadelphia

Draenert K (1988) Forschung und Fortbildung in der Chirurgie des Bewegungsapparates. 2. Zur Praxis der Zementverankerung. Art and Science, Munich

Harris WH, Salzman EW, Athanasoulis C, Waltman AC, De Sanctis RW (1977) Aspirin prophylaxis of venous thrombo-embolism after total hip replacement. N Engl J Med 297:1246—1249

Lee AJC, Ling RSM (1981) Improved cemented techniques. In: Murray DG (ed) Instructional course lectures, vol XXX, Chap 19. pp 407—413, CV Mosby, St Louis Toronto London

Lidgren L, Moller J, Bodelind B (1984) Improved strength of PMMA with vacuum mixing and chilling. Presented at the Swedish Medical Society Meeting, pp 71—79

Pilliar RM, Blackwell R, MacNab J, Cameron HU (1976) Carbon fiber-reinforced bone cement in orthopaedic surgery. J Biomed Mater Res 10:893—906

Rijke AM, Rieger MR (1977) Porous acrylic cement. J Biomed Mater Res 11:373—394

Tepic S, Perren SM (1985) Bone cement preparation with vacuum packaged powder to minimise monomer content and increase strength. In: Draenert K (ed) Die Implantatverankerung. Symposiumsband, Art and Science, Munich, pp 26—27

Ulrich C, Worsdorfer O, Heinrich H (1985) Intra-operative transoesophageale zweidimensionle Echocardiographie bei Huftprothesenimplantaten. In: Draenert K (ed) Die Implantatverankerung, Symposium in Orthopaedie und Chirurgie des Bewegungsapparates, Art and Science, Munich, pp 35—56

Chapter 14

Intrafemoral Pressure in Total Hip Replacement

C. Ulrich and H. Heinrich

Intra-operative trans-oesophageal two-dimensional echocardiography was used to measure the risk of increasing intrafemoral pressure during total hip replacement. This study describes aspects of hip arthroplasty more related to cementation technique than the interface.

Cardiopulmonary Complications

There have been several intra-operative complications related to the cardiopulmonary system since the advent of total hip replacement (Breed 1974; Kallos et al. 1974; Zichner 1972).

The reasons for these complications have been the subject of several investigations. Charnley blamed the cement. Later the post-mortem findings were related to bone marrow, fat and cancellous bone, especially in the lung (Zichner 1972). There is now no doubt that the reason for these complications was raised intramedullary pressure. Kallos and Breed found a raising of the intramedullary pressure at operation when the stem was inserted into the cement. Both authors found that this rise in intramedullary pressure could

be avoided by making a bore hole distal to the prosthesis, and if this was done, there were no cardiopulmonary complications.

Trans-oesophageal Echocardiography

We came to this by chance. A large number of patients were monitored by the anaesthetists using trans-oesophageal echocardiographs (Kremer et al. 1982; Ulrich et al. 1986). A gastroscope with a transducer on the end was inserted into the oesophagus and placed behind the heart so that activity in the right ventricle, the right atrium and the myocardium could be seen (Fig. 14.1). When cement was put into the medullary cavity of the femur, the atrium was filled with contrast colouring which we assumed to be air bubbles.

A second much more impressive phenomenon was a big particle which entered the right atrium when the prosthesis was pressed into the cement (Fig. 14.2). This was accompanied by a significant drop in the end-expiratory pressure of the CO_2. We think that raised intramedullary pressure at operation and the outcome of those emboli from the

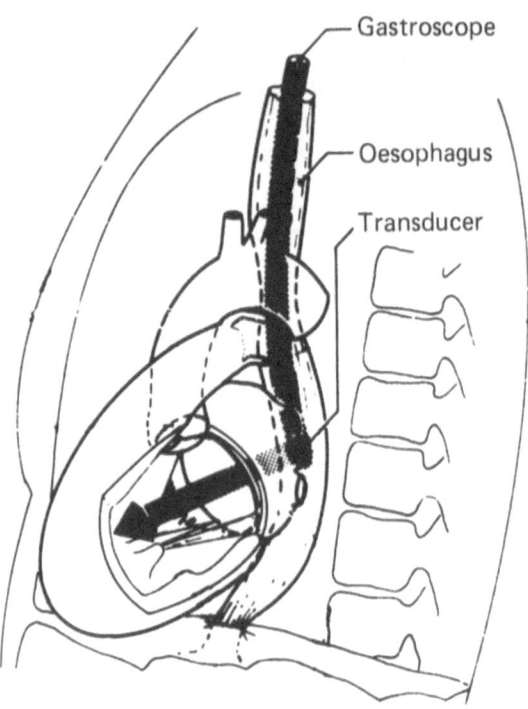

Fig. 14.1. Gastroscope in place behind the heart. The transducer is at the tip of the gastroscope.

femur are responsible for the cardiopulmonary complications that are sometimes seen when a hip replacement is performed. Another research group in Germany did some animal experiments. Air was used to raise the intramedullary pressure and the major vessels were then monitored. A small portion of bone marrow and a big piece of thrombus came out within 2 or 3 seconds after raising the intramedullary pressure.

A patient can be monitored on videotape while a total hip replacement is performed. At the moment the prosthesis is pressed into the cement, the end-expiratory pressure of CO_2 drops and indicates an intrapulmonary embolism.

Two questions arose. First, can embolism during implantation be demonstrated by this means? Secondly, can it be prevented by a venting hole, as Kallos and Breed have shown experimentally?

We had two groups, a control group with 13 patients and the same number with bore holes distal to the tip of the prosthesis. The control group had conventional operations in which a tube was inserted into the medullary cavity while the cement was put in, the tube was then removed and the prosthesis pressed into the cement.

We found air bubbles with contrast colouring in 12 out of 13 patients without a venting hole, and in four patients with a venting hole. Emboli were found in eight out of 13 in the group without a venting hole, and in two patients in the group with a venting hole (Fig. 14.3). These two patients were operated on without the seal that we now use so that the

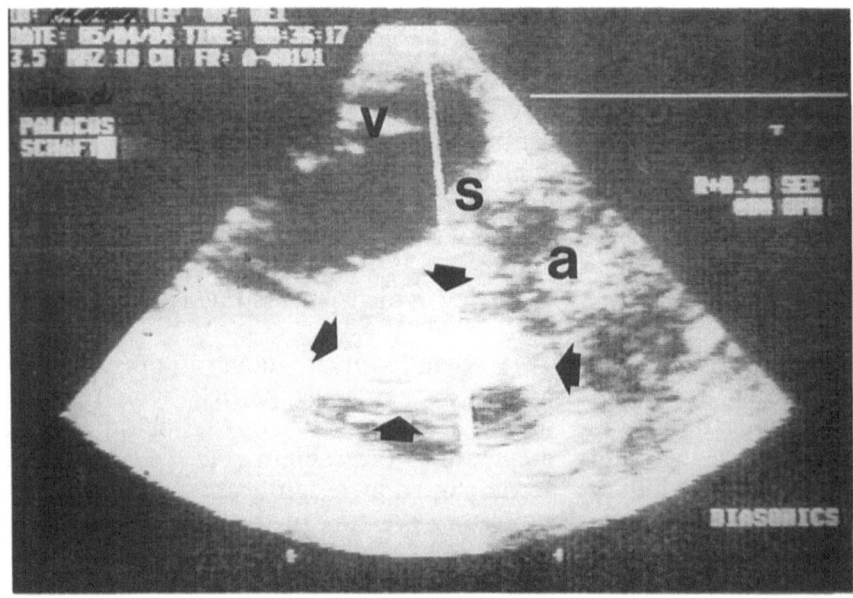

Fig. 14.2. The right atrium is seen in the lower right side of the echocardiographic sector. "Contrast colouring" fills the atrium. *Arrows* indicate the embolus; *a*,atrium; *v*,ventricle; *s*,septum.

Fig. 14.3. Results of the trans-oesophageal 2-D echocardiography of the right atrium and right ventricle during shaft implantation.

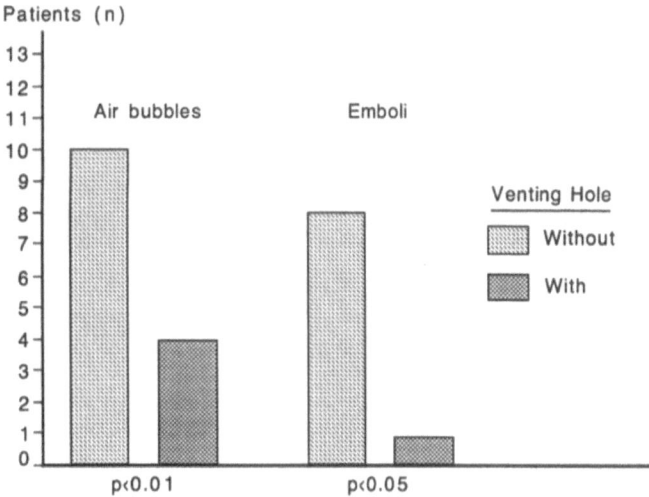

bore hole was closed by cement during the operation. That explains why the intramedullary pressure rate was raised and produced emboli. The end-expiratory pressure of CO_2 dropped in the group without any venting hole. The pressure stayed constant in the group with the venting hole, which means risk to the patient was minimal (Fig. 14.4).

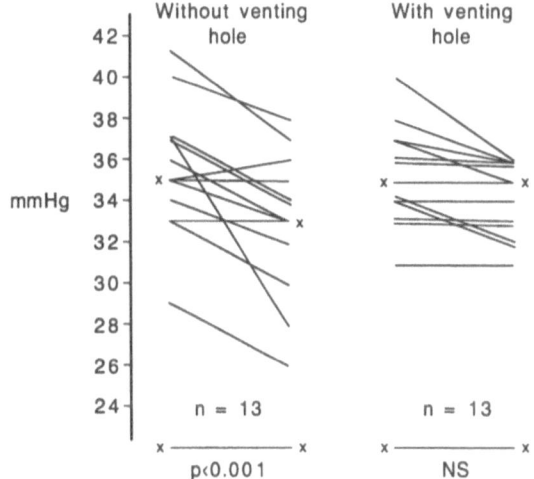

Fig. 14.4. Behaviour of the end-expiratory pCO_2 during shaft implantation.

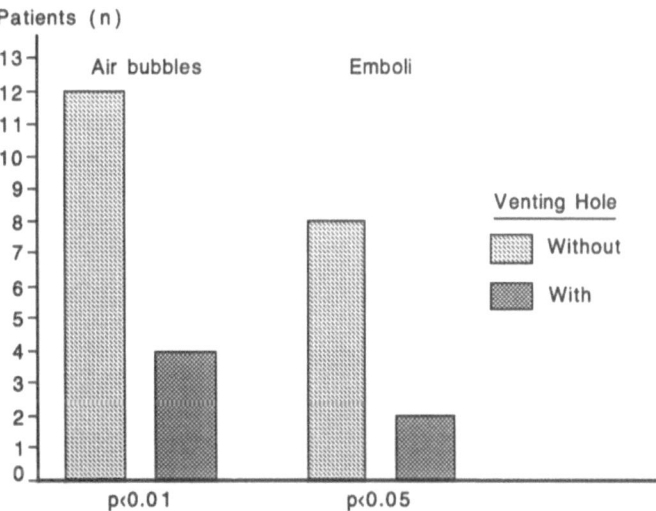

Fig. 14.5. Results of the trans-oesophageal 2-D echocardiography of the right atrium and right ventricle during shaft implantation with an intramedullary seal.

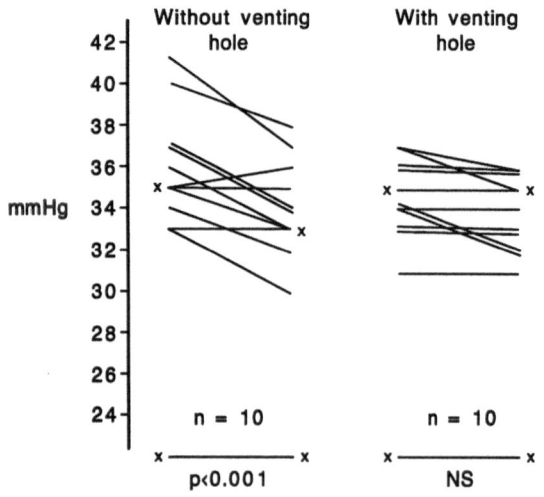

Fig. 14.6. Behaviour of the end-expiratory pCO_2 during shaft implantation using an intramedullary seal.

In the second study we looked for the effect of the intramedullary seal in a conventional operation, and when a distal bore hole was made. A plastic stopper with a central hole was used so that the bone marrow could escape from the proximal to the distal femur. The findings were similar (Figs. 14.5 and 14.6). The air bubbles and the contrast colouring were found in 8 out of every 10 cases in the group without venting holes. That phenomenon can be minimised by putting in a venting hole.

Conclusions

The data from these studies suggest that a distal venting hole together with an intrafemoral plug to stop the cement will improve high pressurisation technique and reduce bone marrow embolism.

Acknowledgement. Dr Ulrich, who presented this paper, is indebted to his co-author for his collaborative work.

References and Further Reading

Breed AL (1974) Experimental production of vascular hypertension, and bone marrow and fat embolism with methylmethacrylate cement. Clin Orthop 102:227—244

Kallos T, Enis JE, Gollan F, Davis JH (1974) Intra-medullary pressure and pulmonary embolism of femoral medullary contents in dogs during insertion of bone cement and a prosthesis. J Bone Joint Surg (Am) 56:1363—1367

Kremer P, Schwartz L, Cahalan MK, Gutman J, Schiller NB (1982) Intra-operative monitoring of left ventricular performance by trans-oesophageal M-mode and 2-d echocardiography (abstract). Am J Cardiol 49:956

Ulrich C, Burri C, Woersdoerfer O, Heinrich H (1986) Intra-operative trans-oesophageal two-dimensional echocardiography in total hip replacement. Arch Orthop Trauma Surg 105:274—278

Zichner L (1972) Embolien aus dem Knochenmarkkanal als Ursache von Sofort-und Spatkomplikationen nach Einsetzen von intramedularen Femurkopfendoprothesen mit Polymethylmethacrylat. Helv Chir Acta 39:717—723

Part III

Bone

Chapter 15

Bone: The Architecture of Bone and How it is Influenced by External Loading

L.E. Lanyon

Everyone will agree that one of the principal influences on whether a prosthetic fixation succeeds or not is the situation at the interface. The interface will react in many ways according to its specific characteristics. However, the interface is also the perimeter of the bone architecture whose position is determined by another set of influences which result from its role as a bearer of functional loads. The data presented relate to the functional determinants of bone architecture.

Functional Load Bearing

Bone architecture is obviously influenced from a number of sources, but it is important to remember that the only functional requirement for any particular shape is that related to functional load bearing. Much is written of the hormonal influence on bone architecture, but there are no means by which hormones can get the necessary feedback to control bone architecture. They set the milieu in which mechanically related processes operate, and thus may enhance or impede the speed and balance of the modelling and remodelling processes. However, the only influences which contain the necessary information on the adequacy of bone architecture in relation

to its functional role come from functional load bearing itself.

In a 40-week-old foetus, the influences are already becoming apparent (Fig. 15.1) (Rodriguez et al. 1988). Comparison of para-

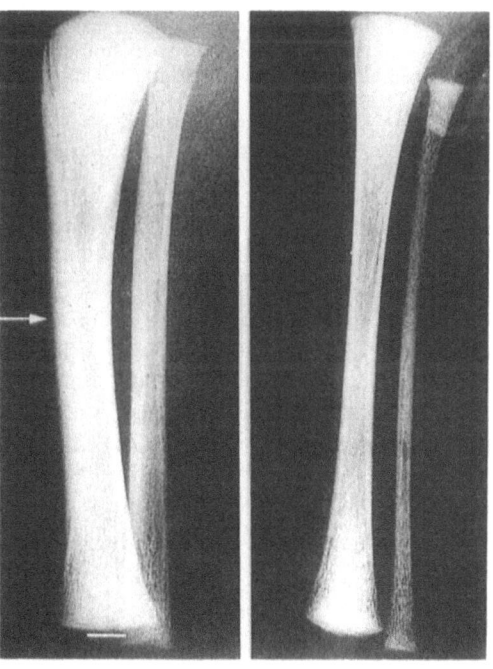

Fig. 15.1. Radiographs of left and right tibiae and fibulae in 40-week-old human foetus with unilateral paralysis showing the effects of lack of muscle loading in utero (Rodriguez et al. 1988).

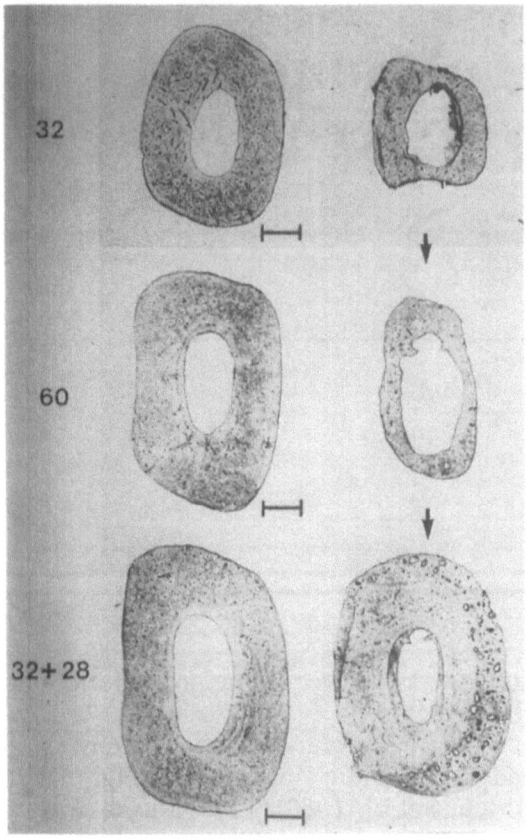

Fig. 15.2. Showing the failure of the dog metacarpal bones to develop in the absence of loading for 32 or 60 weeks and the reversibility of this process after 32-week immobilisation and 28-week remobilisation (Jaworski and Uhthoff 1986).

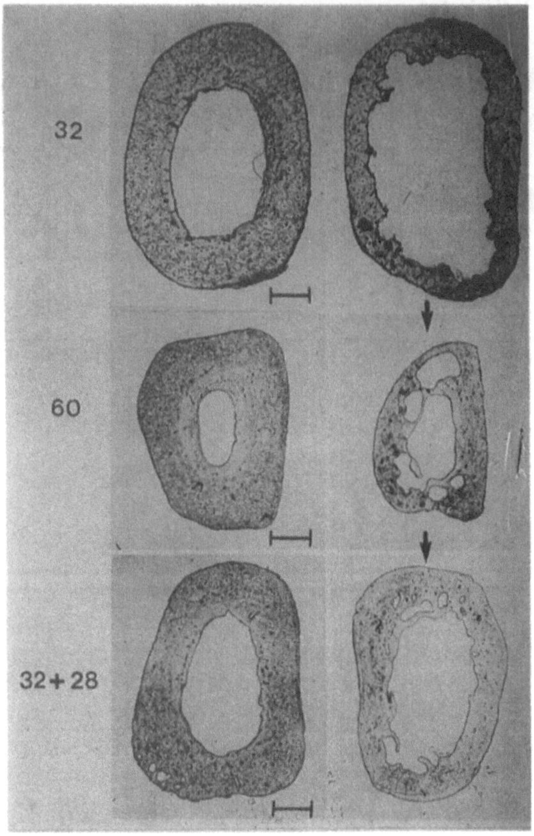

Fig. 15.3. Showing the loss of bone after 32- and 60-week immobilisation followed by partial restitution after 28 weeks remobilisation (Jaworski and Uhthoff 1986).

lysed limbs with normal limbs shows that the features on which functional load-bearing competence of the skeleton depends — that is, the bone's girth, its curvature, its cortical thickness and the amount of its cancellous bone — can be seen to require functional load bearing for their development. In experiments (Uhthoff and Jaworski 1978), Jaworski and Uhthoff (1986) showed the effect in growing Beagle dogs of a bone that is normally functionally loaded and one that is deprived of functional loading. In the growing situation the normal architecture of the bone, its shape, size and mass, are not achieved unless the stimulus of growth is supplemented by the stimulus of functional load bearing.

This lack of development is reversible if function is restored. Thirty-two weeks of disuse followed by 28 weeks of restored function resulted in new bone formation as the functionally deprived bone caught up with the normal bone (Fig. 15.2). So it can be seen that functional load bearing is necessary not only to achieve but also to maintain normal bone architecture (Fig. 15.3).

Presumably the objective of the relationship between bone architecture and bone loading is to make sure that the loads to which the bone is subjected are not so great that they cause a level of damage which cannot be repaired by normal remodelling. Whatever loading-related parameter it is that affects the cells the changes that they can make are to material properties of the bone, its mass and its architecture. Fortuitously, the product of load, architecture, material properties and mass is the strain or deformation that occurs within the bone as a result of functional load

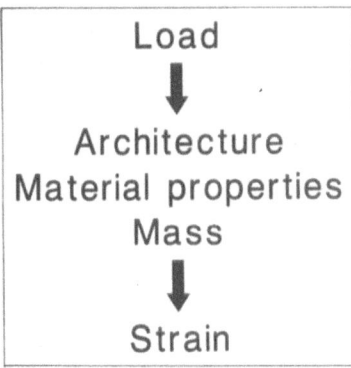

Fig. 15.4.

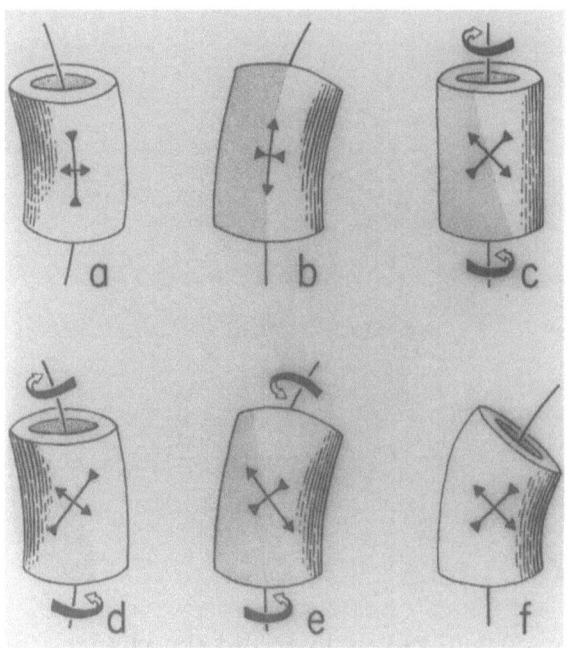

Fig. 15.6. Pictorial illustrations of surface principal strains under difficult loading conditions (Lanyon and Bourn 1979).

bearing (Fig. 15.4). Strain, unlike stress or applied load, is a parameter to which bone cells respond. Stress, of course, is a figment of engineers' imaginations designed purely to explain strain.

A strain pattern, even under a very simple loading situation, is fairly complex. One of the simplest loading situations that could be imagined is shown in Fig. 15.5 where a block of material has a compressive load applied to it. The material squashes more or less in the direction of the compressive load but in the direction opposite to that it stretches. The directions of principal compression and principal tension are always at right angles to one another. By definition there is no shear stress component in these principal directions, but

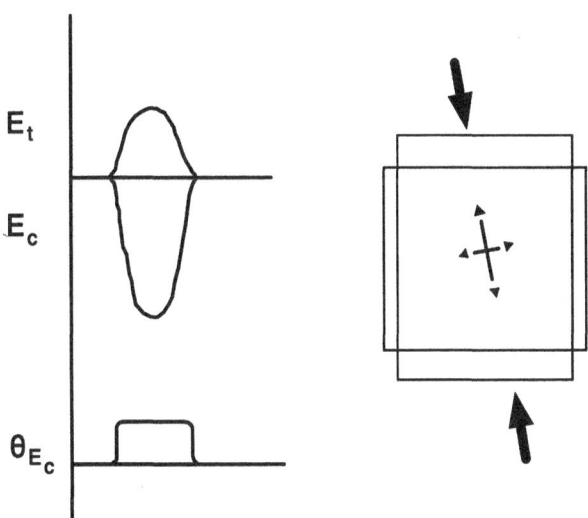

Fig. 15.5. Pictorial and graphical representations of the principal tension (Et) and principal compression (Ec) strains and the angle of these to any designated axis.

in any other direction there is a component of compression, tension and shear. So at any one particular location on a piece of bone, even in one plane, the parameters that need to be known may be the tension strain component, the compression strain component and the angle that these make to whatever axis it is that interests you. Change in principal tension, principal compression and angle can be plotted as shown in Fig. 15.6 (Lanyon and Bourn 1979).

For us to be able to examine the mechanism whereby bone adapts its architecture to the normal functional loading of everyday activities or to the changed functional loading in a prosthesis situation, we thought it was important to be able to sample the same strain information which is received by the bone cells themselves.

For a number of years we have done this by attaching strain gauges, all measuring strain from a single location, to bone surfaces in vivo (Lanyon and Smith 1970). We even managed to induce one member of the orthopaedic profession to have a strain gauge attached to his tibia (Lanyon et al. 1973). As far as I know, this is still the only documented case of assessing strain from the human situa-

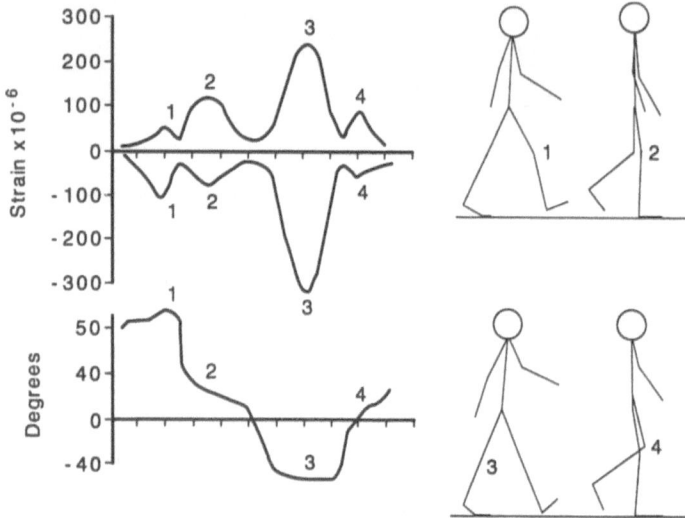

Fig. 15.7. Graphical representation of the changing principal tension and principal compression strains and their angle to the tibia long axis during human locomotion (Lanyon et al. 1973).

tion. It allowed us to record the changes in tension and compression strains and the change in the angle of these strains during locomotion. It also allowed us to demonstrate that the strains in the human skeleton were exactly the same as all the strains that we measure in a great many locations in animal experiments, which is, of course, what one should have expected in the first place (Fig. 15.7).

We recorded strains in a great range of locations within the animal skeleton, and we thought originally that there was going to be some uniform strain value, 827 for instance, throughout the skeleton. In fact we found that the strain situation throughout the skeleton was certainly within a definable strain range, but the distribution and magnitude of the strain varied very greatly in different locations. However, it appeared to be critical in each location. Thus, a strain situation that was perfectly appropriate for the mid-shaft of the humerus would produce tremendous adaptive responses if applied to the mid-shaft of the tibia, and vice versa. So it seems there is a site-specific situation, sensitive to strain, but there is no uniform strain parameter that we have identified which bone modelling and remodelling is directed to achieve.

To try to identify the actual features of the strain environment to which the bone responded, we produced a functionally isolated externally loaded bone preparation (Lanyon and Rubin 1984). The bone was the

ulna in the wing of birds. We have used both chickens and turkeys but preferred the latter. The preparation involves making osteotomies both proximally and distally to produce a functionally unloaded situation, but one which is externally loadable through Steinmann pins which penetrate through the ends of the bone and caps which cover its end. The pins are engaged with a loading machine. We can produce in vivo a fairly large (110-mm long) organ culture, kept alive by the animal, in which we could determine both the load and strain environments, and, with the aid of fluorochrome labelling, follow the remodelling.

The modelling and remodelling within the bone were affected profoundly by varying the external load as can be seen in Fig. 15.8 (Rubin and Lanyon 1985). All that is necessary for the quiescent remodelling state in the functional bone to be converted to the active phase of disuse osteoporosis is to remove functional load bearing. As soon as it is removed, tremendous activation of remodelling occurs together with endosteal resorption and intracortical invasion of cutting cones. This is classic disuse osteoporosis and illustrates the first responsibility of normal functional loading, which is to preserve what is there and constantly counter the hormone-mediated resorption that would occur if functional load bearing was reduced.

If we applied an external dynamic load instead of disuse for short periods each day,

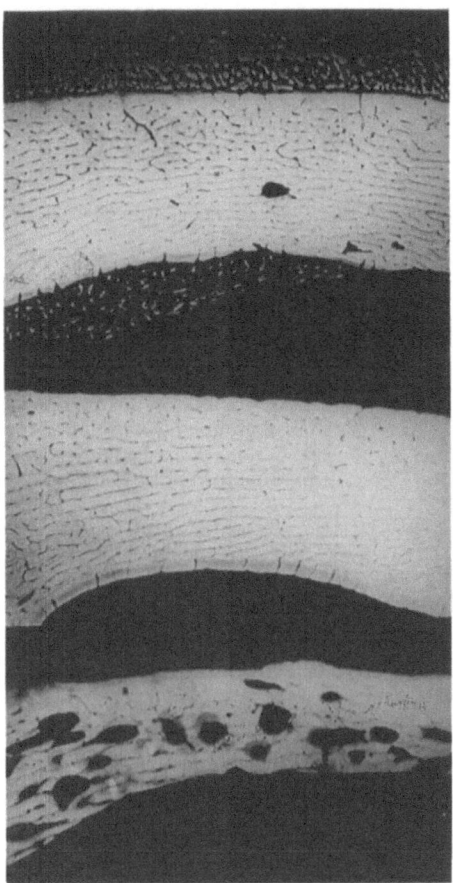

Fig. 15.8. *Middle*: the quiescent remodelling state of an intact adult turkey ulna. *Below*: the active stage of resorptive modelling and remodelling characteristic of the active phase of disuse osteoporosis. *Top*: the effect of disuse interrupted by a single short daily period of loading showing inhibition of resorption and a strain-magnitude-related increase in periosteal and endosteal new bone formation (Rubin and Lanyon 1985).

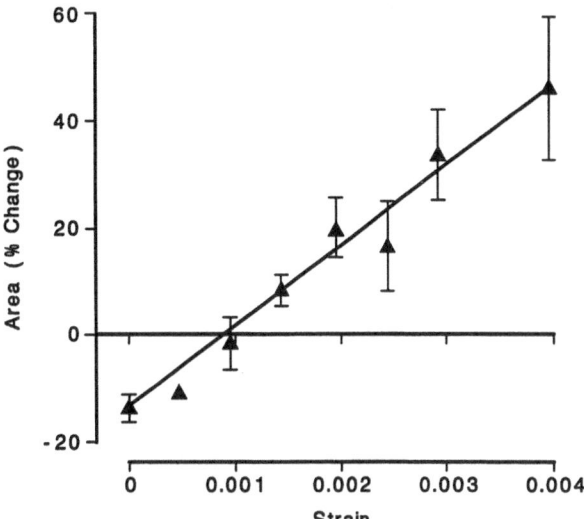

Fig. 15.9. The dose—response relationship between peak strain magnitude and changes in bone area illustrated in Fig. 15.7 (Rubin and Lanyon 1985).

we could produce a completely different situation in which resorptive remodelling was prevented and the integrity and thickness of the cortex prevented. In addition, new bone formation on both periosteal and endosteal cortices could be induced. When the change in bone cross-sectional area is plotted against the peak applied strain, there is a satisfactory dose—response curve showing that when strains are low or absent, disuse osteoporosis occurs, but as the strain level increases, so does the osteogenic effect (Fig. 15.9).

This result is interesting because although at the higher strains we applied there was a substantial increase in bone mass, the actual peak strains applied were no larger than the bird itself could produce and could automatically engender in the intact state by flapping its wings.

This raises the question "What is the difference between the strain situation when we apply it and get massive osteogenic adaptation, and when the animal applies it, and produces no change?" We suppose that the difference is the strain distribution. At peak strain, when the animal itself flaps its wings, a portion of the bone is in compression and a portion in tension. When we applied external loading, the peak strains were the same but their distribution was different and so different areas were in tension or compression. We assume that it is this abnormal strain distribution that forms the basis of the adaptive stimulus. Because they are best placed to make that sort of assessment, we assume that it is the interconnecting osteocyte population that distinguishes between normal and abnormal strain magnitudes and distribution.

That presupposes that there will be a whole series of different load or strain response curves for different strain distributions. Some strain distributions are very foreign to the bone and would result in a very large magnitude-related response, and some are very close to normal bone distribution producing a very small magnitude-related response. This leads inevitably to the hypothesis that if you get exactly the right normal strain distribution

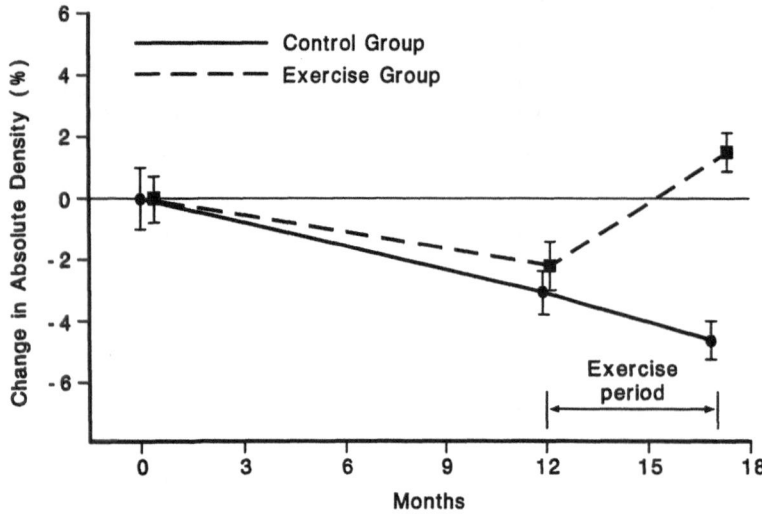

Fig. 15.10. Percentage changes of the mean bone density in the exercise and control groups 1 year prior to the exercise period, at its beginning and at its end (Simkin et al. 1987).

there is no adaptive change (Lanyon 1987). This means we have an error driven system which is totally different from the situation we originlly envisaged in which we thought that the predominant functional influence would produce the predominant functional effect.

Is Functional Adaptation Error Driven?

What we are suggesting now is that it may not be normal coordinated functional loading that causes the architectural changes or maintains the adaptations, but that there are responses to everyday loading accidents. That, of course, distinguishes the response of functional adaptation from that of damage, because fatigue damage, creep and such like are determined by many repetititions of function activity.

This hypothesis is supported to some extent by the finding that in the externally loadable system where disuse is interrupted by a period of artificial loading, there is a tremendous adaptive response from very few loading cycles per day (Rubin and Lanyon 1984). This system, therefore, is very responsive to a very few strain reversals which would be consistent with its search for comparatively few error signals. Perceiving errors and adapting to them rather than constantly monitoring

features of repetitive strain cycles derived from everyday activities might be the sensible way to design an adaptive system.

One of the fields in which we are trying to work out the mechanisms behind loading-related changes in bone architecture is, of course, in relation to osteoporosis. The elderly female patients in Professor Simpkin's osteoporosis clinic were asked to do a number of loading exercises which were abnormal to them (Simpkin et al. 1987). These Israeli grandmothers, on whom he was measuring bone at the wrist, were persuaded to do abnormal exercises with their wrists which involved both arm wrestling and hanging from ladders, which he assumed was something they did not normally do! It was very satisfactory to see that in those who performed these exercises the bone mass at their wrists started to rise immediately afterwards, while the bone mass at the wrist in the controls who carried on their lives normally continued to decline (Fig. 15.10).

In the situation of prosthetic implants the host bone is condemned to a totally different strain situation in terms of strain distribution and probably strain magnitude from that to which it is normally accustomed. In relation to error signals this may lead to adaptive remodelling. Since the host sites are condemned to a situation of abnormal strain distributions the secret of long-term prosthetic fixation must be to try to achieve beneficial strain situations. If the beneficial error can be achieved this will enhance rather than detract from the quality of prosthetic fixation.

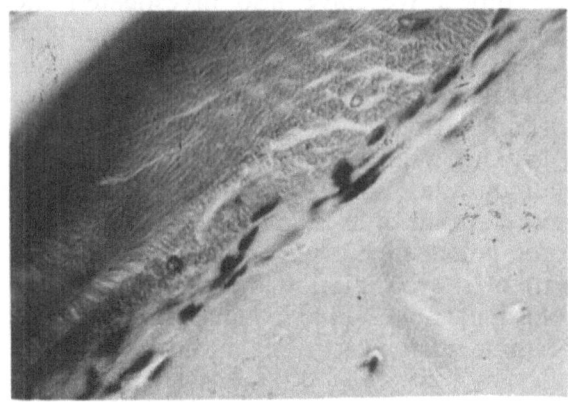 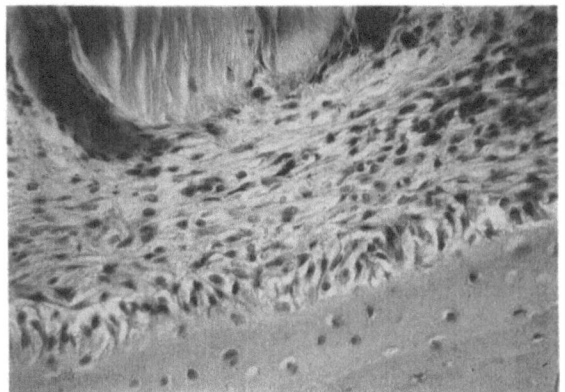

a b

Fig. 15.11a. Quiescent periosteal surface in an adult turkey ulna. **b** A similar bone subjected to a single "osteogenic" period of loading 5 days previously (Pead et al. 1988a).

When we recorded the strain situation at the femoral calcar in situations with and without total hip replacement, there was a reduction in the normal degree of compression in the implanted femur which was accompanied by calcar resorption (Lanyon et al. 1981). We think we answered the question as to whether it was caused by disuse or not by making exactly the same insertion in a case where there was a sciatic neurectomy. In this situation there was massive loosening and massive endosteal resorption which might have been expected from an unloaded situation but the shell of calcar was nicely preserved. This implies that there is a difference between uniformly reduced lack of function, with presumably no abnormal strain distribution, and the actively driven remodelling situation where there is an abnormal error signal as in the calcar when it was being functionally loaded.

The Osteogenic Response to a Single Period of Loading

Functional loading may influence bone architecture in various ways. When repeated daily, only a very few loading cycles are necessary to produce a sustained osteogenic response. It is obviously interesting to know what happens if just one loading incident is given on a particular day and then not repeated.

Figure 15.11a shows normal quiescent periosteum. One 5-minute period of intermittent loading transfers that quiescent periosteum into one in which new bone is being formed (Fig. 15.11b). This means that the whole cascade of events involving recruitment of periosteoblasts, cell division, matrix synthesis, ossification and so on, can be set in motion by one osteogenic event on one particular day (Pead et al. 1988a). Sustained remodelling, therefore, is presumably the cumulative effect of a number of these events occurring one after the other, as a result of repeated exposure to different loading situations.

Loading-Related Activity in Osteocytes

If our hypothesis was correct and osteocytes are the strain-sensitive cells, then we should be able to assess their responsiveness to mechanical strain. In a number of animals we gave tritiated uridine at the same time as the loading event. The number of osteocytes taking up tritiated uridine, which we assume to be making RNA, increased by a factor of 12 in a loaded compared with an unloaded bone (Fig. 15.12) (Pead et al. 1988b).

There is a similar loading-related increase in the number of closing osteocytes — G6PD enzyme activity, only 5 minutes after a period

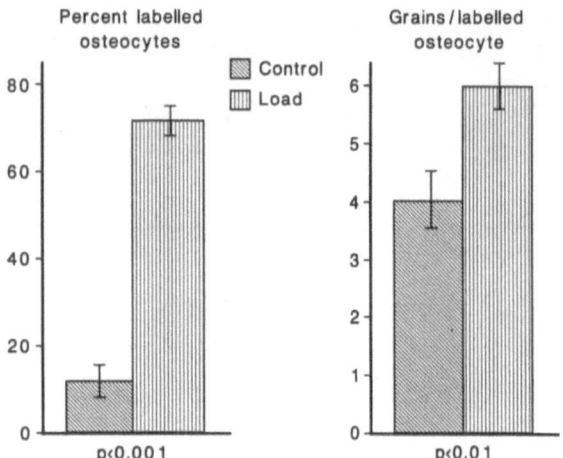

Fig. 15.12. The percentage of osteocytes taking up a 3H-uridine label and the number of grams per osteocyte in bones which had been loaded 24 hours previously compared with those which had not (Pead et al. 1988b).

of loading (Skerry et al. 1989). Not only do we get an increase but it is a strain-related increase (Fig. 15.13). So it does seem possible, if not probable, that osteocytes adapt their biochemical response to loading immediately afterwards, and this may be early evidence of the osteogenic response that can only be seen as new bone formation some 5 days later.

Strain Memory

Although I have concentrated on the bone cells' ability to respond to transient strains in

"real time", we have also investigated the possibility that there is a strain memory in bone tissue. Osteocytes are surrounded, even in adult cortical bone, by proteoglycan molecules. These molecules can be orientated by looking with the electron microscope at the molecules themselves, or by looking at their Alcian-blue-induced biorefringence under polarised light. A quantitative assessment of the orientation of the proteoglycan molecule is obtained by comparing the degree of biorefringence in loaded and unloaded sections.

If two sections are taken, one from an unloaded and one from a loaded bone, there is a difference in the degree of biorefringence between the loaded and unloaded immediately after loading. This implies that the dynamic loading has moved the proteoglycan molecules so that more of them are oriented in the direction of the collagen. That of itself might not have been too interesting, but 24 hours later in the absence of further loading the degree of increased orientation was still present and had not been eliminated (Skerry et al. 1988) (Fig. 15.14). This means that there is a mechanism, whether or not it is used, for actually providing a record of transient strain experience. The effects of transient strains can be captured within the orientation of proteoglycan molecules, and then retained for a period of some hours. So it is possible that although the osteocytes are quickly "awakened" after a period of being loaded, they are "informed" as to the strain distribution or the pattern of strain within the

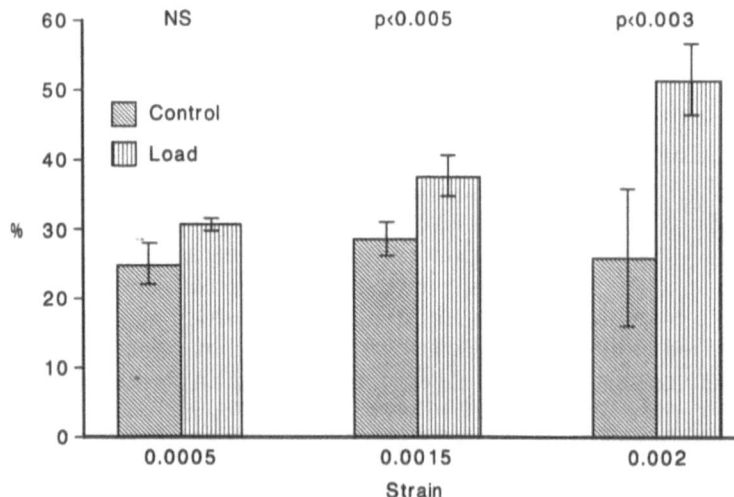

Fig. 15.13. The strain-magnitude-related change in the activity of the enzyme glucose-6-phosphate dehydrogenase in osteocytes strained in vivo at different levels 5 minutes after loading.

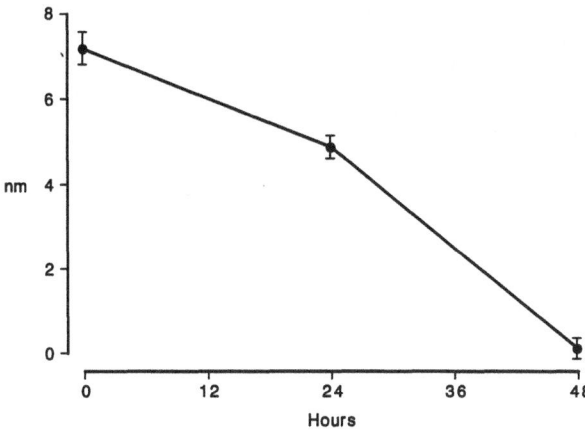

Fig. 15.14. Optical path difference (nm) between Alcian-blue-stained sections from non-loaded control bones showing the induced birefringence immediately after loading declining over the next 48 hours.

bone tissue by looking at the proteoglycan molecules, which are found throughout the matrix and are intimately connected with the bone cell membrane.

We have seen this loading-related bio-refringence in every place that we have looked. Interestingly, Ferris and his co-workers (1987) have shown that in human osteoporotic bone, one of its features is a lack of proteoglycan orientation (Fig. 15.5). Unfortunately, we do not know whether that simply means that human osteoporotics have been inactive for some time and what we are seeing is a reflection of disease or whether

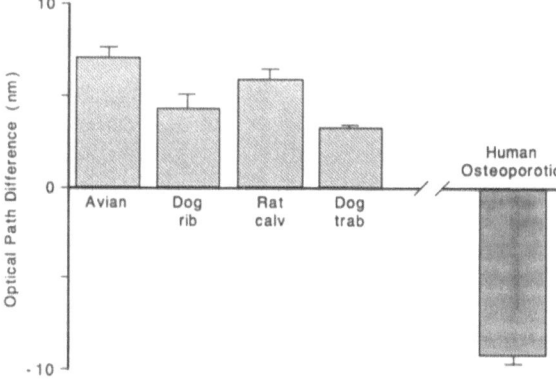

Fig. 15.15. The optical path difference in bones which had or had not been loaded showing the positive-loading-related increase compared with the decreased level in clinical human osteoporotics compared with normal (Ferris et al. 1987).

this is orientation illustrating a deficiency in the mechanism by which loading affects the biological system. If this were the case the osteoporosis could be a reflection of an inadequacy in the way in which loading events are captured and subsequently referred to.

Certainly proteoglycan molecules actually cross with cell membranes in other cell systems and attach themselves to the cyto-skeleton. It is possible that loading and strain in the matrix pulls levers outside and rings bells inside the cell. If you have the eye of faith, as every investigator must have, one can assume an association between the active elements of the cyto-skeleton and the proteoglycan.

Summary

The continuing ability of the skeleton to resist fracture depends upon maintaining at each location sufficient bone tissue with adequate material properties, appropriately placed to withstand applied loads.

At each location the product of bone mass, architecture, material properties and applied load is the dynamic strain situation within the tissue. Functional strains, therefore, carry the information necessary for the feedback control of bone architecture in relation to load-bearing requirement. Such information cannot be provided by systemic influences.

The importance of functional strains in the control of bone architecture is supported by the findings of immediate strain-related changes in the activity of resident bone cells, and long-term strain-related changes in modelling and remodelling behaviour.

We propose that the modelling and remodelling which actually determine bone architecture at each location are primarily the result of the balance between potentially competing and complementary influences derived from local strain-related factors and the local effects of systemic hormones.

With an implant, the bones' mechanical situation is disturbed, irrevocably so in the case of prostheses. This change in strain environment will engender adaptive remodelling which will, in turn, be perturbed by local effects derived from the particular events at the bone—implant interface. The objective of good implant design must be to ensure a strain environment within the host bone tissue and at the interface that is compatible with a modelling and remodelling situation which increases rather than decreases the stability of the fixation.

References and Further Reading

Ferris BD, Klenerman L, Dodds RA, Bitensky L, Chayen J (1987) Altered organisation of non-collagenous bone matrix in osteoporosis. Bone 8:285—288

Jaworski ZFG, Uhthoff HK (1986) Reversibility of non traumatic disuse osteoporosis during its active phase. Bone 7:431—439

Lanyon LE (1987) Functional strain in bone tissue as an objective, and controlling stimulus for adaptive bone remodelling. J Biomech 20:1083—1093

Lanyon LE, Bourn S (1979) The influence of mechanical function on the development and remodelling of the tibia. J Bone Joint Surg (Am) 61:263—273

Lanyon LE, Rubin CT (1984) Static versus dynamic loads as an influence on bone remodelling. J Biomech 17:892—905

Lanyon LE, Smith RN (1970) Bone strain in the tibia during normal quadrupedal locomotion. Acta Orthop Scand 41:238—248

Lanyon LE, Hampson WEJ, Goodship AE, Shah JS (1973) Bone deformation recorded in vivo from strain gauges attached to the human tibial shaft. Acta Orthop Scand 46:256—268

Lanyon LE, Paul IL, Rubin CT, Thrasher EL, DeLaura R, Rose RM, Radin EL (1981) In vivo strain measurements from bone and prosthesis following total hip replacement. An experimental study in sheep. J Bone Joint Surg (Am) 63:989—1000

Pead MJ, Skerry TM, Lanyon LE (1988a) Direct transformation from quiescence to bone formation in the adult periosteum following a single brief period of bone loading. J Bone Min Res 3:647—656

Pead MJ, Suswillo R, Skerry TM, Vedi S, Lanyon LE (1988b) Increased 3H-uridine levels in osteocytes following a single short period of dynamic bone loading in vivo. Calcif Tissue Int 43:92—96

Rodriguez JI, Palacios J, Garcia-Alix A, Pastor I, Paniagua R (1988) Effects of immobilisation on foetal bone development. A morphometric study in newborns with congenital neuro muscular diseases with intra-uterine onset. Calcif Tissue Int 43:335—339

Rubin CT, Lanyon LE (1984) Regulation of bone formation by applied dynamic loads. J Bone Joint Surg (Am) 66:397—402

Rubin CT, Lanyon LE (1985) Regulation of bone mass by mechanical strain magnitude. Calcif Tissue Int 37:411—417

Simkin A, Ayalon J, Leichter I (1987) Increased trabecular bone density due to bone loading exercises in postmenopausal osteoporotic women. Calcif Tissue Int 40:59—63

Skerry TM, Bitensky L, Chayen J, Lanyon LE (1988) Loading-related reorientation of bone proteoglycan in vivo. Strain memory in bone tissue? J Orthop Res 6:547—551

Skerry TM, Bitensky L, Chayen J, Lanyon LE (1989) Early strain related changes in enzyme activity in osteocytes following bone loading in vivo. J Bone Min Res 4:783—788

Skerry TM, Suswillo R, El Haj AJ, Ali NN, Dodils RA, Lanyon LE (1990) Load induced proteoglycan orientation in bone tissue in vivo and in vitro. Calcif Tissue Int (in press)

Uhthoff HK, Jaworski ZFG (1978) Bone loss in response to long term immobilisation. J Bone Joint Surg (Br) 60:420—429

Discussion

The Chairman - **Professor Fitzgerald**

The Panel - **Dr Lee**
- **Dr Draenert**
- **Dr Ulrich**
- **Professor Lanyon**

Mr Compton: Dr Lee demonstrated extrusion downwards and sideways into gaps. May I ask him whether there was any upwards extrusion out of the gun?

Dr Lee: Downward extrusion was gross and easily visible. We were not able with our experimental set-up to measure whether it had moved upwards or not. It is perfectly possible that it could have extruded upwards as well as downwards and sideways.

Professor Fitzgerald: The specimens seemed to go outward but not to have the same shape as the V-cuts that you had in your taper control.

Dr Lee: This is quite true. The Simplex bone cement specimens only partially filled the gaps, but I suspect that is simply a matter of time. The other bone cement that Freeman and his group have put together, which is a different chemical, did in fact creep very much more readily and entirely filled the gaps. I think it is purely a time-related phenomenon.

Mr Northmore-Ball: May I ask Professor Lanyon to discuss the changes in size and diameter of the intramedullary canal which occur with ageing in long bones?

Professor Lanyon: As you know, it has been suggested that the periosteal expansion might be a compensation for the hormone-mediated endosteal resorption. It is possible that the minimal amount of strain required to stimulate remodelling occurs endosteally in bones that are primarily bent, but no one really knows.

Mr Macdonald: May I ask Dr Lee what would be the effect on his creep experiments of stem-to-cement adhesion?

Dr Lee: I think it could be profound. Plainly bone cement, like every other material, will react to the three-dimensional stress pattern that is put upon it. I believe the stem-to-cement interface does profoundly affect the stress and consequently strain distribution within the cement mantle.

Professor Solomon: I have a question for Professor Lanyon. In the experiment in which you found periosteal bone formation after a fairly short period, did the bone thickening occur only at the periosteal surface, or was there a change in the internal trabecular architecture as well after such short periods?

Professor Lanyon: If you mean the single period of loading, we were interested only within the first week. If loading is not continued, the new bone that was formed is removed again and the classical effects of disuse osteoporosis occur. One period of loading results in an osteogenic expression, but from the second day disuse osteoporosis has taken over. Sustained adaptive change would only be seen if you gave the foreign strain situation daily for a period of weeks. Then I would confidently expect that there would be changes in the trabecular architecture as well.

Professor Solomon: I was interested in the fact that the change occurred so very quickly, whereas if it was influencing normal physiological bone turnover it would take at least 3 months.

Professor Lanyon: There are two points. First, it is initiated very early. You have to remember also with these experiments that we have created an artificial condition. We have a disuse situation, in other words darkness then a flash of instructions then darkness again. That may be very different from a situation where there is a constant stream of functional input and a few interspaced peaks. If you fall downstairs, for instance, your memory is full of instructions to remodel by the time you reach the bottom of the stairs. By the time you have staggered off, the nurse has been nice to you and you have walked to your car and got in, however, all that memory may have been degraded by subsequent information telling the skeleton that things are all right. Although disuse experiments like this allow us to look at the mechanism, they are not necessarily immediately transposable to a situation where you have both normal and abnormal instructions coming in at the same time.

Dr Draenert: Do you think that unloaded osteoporotic bone, if later loaded, will regain the same shape morphologically? Is there any difference between the bone in the diaphysis and in the metaphysis where you have cancellous bone?

Professor Lanyon: No. It will never regain exactly the same shape, because nothing ever does. Theoretically, I see no reason why it should not. Certainly if you leave bone for a long period of time there are some people who say it is no longer responsive, but we have some experimental evidence to the contrary. It is impossible, however, to reproduce the experiment exactly. In Jaworski's experiment a dog was in a plaster cast for 40 to 60 weeks. When the cast was taken off there were changes, but there is no information as to whether normal functional loading was ever restored. This may relate to the fact that you do not get normal bone architecture.

Professor Fitzgerald: Dr Ulrich, if you have introduced a stress riser distal to your composite femoral bone cement proximal femur, there may be fractures. Have you seen fractures and are you concerned about them?

Dr Ulrich: No, we have never seen fractures through this hole, since we do it only on the lateral cortex. When it was first tried in Germany in 1974, they had a failure like that in Mains, but they perforated the medial cortex as well, and they did not take any bony seal as we do now. The bony seal means healing of the holes within 6 to 8 weeks so there is no more stress concentration. A plastic plug or a plug made out of polymethylmethacrylate will fail, but if you use bone from either the femoral head or intertrochanteric area as a plug the holes will heal completely. Draenert did the original experiments with primates.

Mr Atkins: Dr Lee, are the physical properties of cement related to its molecular changes when it is immersed in fat?

Dr Lee: If you analysed normal bone cement 2 hours after it had apparently polymerised and hardened, you would find that 85% of the available monomer had polymerised. In consequence, after hardening there is a post-polymerisation process that goes on in an exponential way until, after a few weeks, all but 3% of the monomer will have polymerised. That is one ageing phenomenon that will continue. Liquid like saline and certainly the fats will then act as plasticisers within the polymeric matrix. Again, that is a reasonably well known and experienced phenomenon. So we have a combination of post-hardening polymerisation and plasticisation going on which does account for these changes. I think there would be very little change, although we have not done long-term tests. I have some specimens in the oven at 37°C and we shall soon be able to do 6-month and 1-year tests, but as yet I do not know the difference between 6 weeks and 6 months. I would expect it to be very small.

Mr Ling: What does Dr Ulrich feel about the effect of lavage on the femoral canal, because I believe that Waddell and his colleagues in Toronto did some work with dogs in which they showed that these pulmonary effects could be enormously reduced simply by lavaging the canal before cement insertion, and of course, introducing it retrograde. There was a big difference between the lavaged and unlavaged canals. Do you have any comment about that?

Dr Ulrich: It is true that lavage of the canal significantly reduces the embolism, but you cannot reduce the intramedullary rise of pressure by this method.

Mr Ling: I think there was some other experimental work done in cats in this country in which cardiovascular changes were related to a neurological mechanism initiated by a very high increase in intramedullary pressure as opposed to embolisation.

Dr Ulrich: That phenomenon was found but I do not know of any publication relating to human beings.

Dr Lee: Lavage is obviously a phenomenon before insertion of bone cement. You also mentioned that insertion of the implant was significant. Does the rate of insertion have any effect? This will govern to an extent the behaviour of bone cement, which is viscoplastic, and the pressures generated within it. Do you know whether there is a significant difference if you push it in as quickly as you can or if you just allow it to float into the canal?

Dr Ulrich: We have not measured the rate of insertion, but feel that slow input of the prosthesis may cause a slow rise in the intramedullary pressure which may be significant.

Mr Miles: Dr Lee's tests have shown significant changes in the properties of cement during the early post-operative period. Has he any comment to make on the post-operative rehabilitation regime for the patient? When should weight bearing be allowed, because clearly the movement of cement early on may predispose to later problems?

Dr Lee: Yes, the early effects of cement properties could have a considerable influence on post-operative care. As I am an engineer, perhaps my colleague, Robin Ling, would be prepared to comment on the significance of the creep results that we are getting with post-operative care.

Mr Ling: From 1970 to 1975, patients were allowed to weight bear straight after surgery but subsidence was seen in a high proportion of those cases. Now patients are not encouraged to weight bear in the same way unless they are very old. We still see the phenomenon of slight subsidence of the stem within the cement. I am unconvinced that the femur needs protection, but the socket does.

Professor Sloof: Dr Draenert, I think it is quite evident with regard to the micro-interlock, that the bone cement must penetrate into the trabecular structures. We have now seen this very sophisticated vacuum system. How is it possible to control the depths?

Dr Draenert: It is absolutely controlled by the position of the cannula. If the cement will reach the open end of the cannula, the vacuum has no further effect. It might be of special interest for the ilium. In planning the operation you have to plan the position of the cannula.

Part IV

Miscellaneous

Chapter 16

Endosteal Lysis

R.S.M. Ling, P.P. Anthony, G.A. Gie and C.R. Howie

Localised endosteal bone lysis in the femur in relation to femoral components that are not obviously loose has not, as yet, been satisfactorily explained.

Several authors have suggested that such lesions often start in areas where there is apparently close contact between the metal of the stem and the endosteal surface of the femur (Carlsson et al. 1983; Huddleston 1988). Others have not specifically commented on this point, but X-rays shown in their papers demonstrate it nevertheless (Jasty et al. 1986; Lombardi et al. 1989).

We have explored directly areas of localised lysis in three cases. We found the lytic area in each to be associated with a defect in the cement mantle, so that the metal surface of the stem was in direct communication through the mantle defect with the lytic area. One explanation for the lysis immediately comes to mind in relation to the stem—cement interface in the findings of Fornasier and Cameron (1976). These authors showed in a postmortem study of five curved, collared Muller femoral components (in which the acrylic cement was dissolved out by immersing the femora in methyl-methacrylate monomer) that there was invariably present a fibrous layer, up to 100 μm in thickness, that completely separated the stem from the cement. Our findings were the same in a similar study of three matt-surfaced Exeter femoral components. Thus, a fibrous layer was found in all eight cases in which this membrane was sought. Statistically, it is likely to be present, therefore, in not less than 75% of cases unless special measures are taken to prevent its formation (Morgan, 1989, personal communication). The cause of the void between stem and cement in which the fibrous layer forms is uncertain. It probably depends upon differential volumetric changes between stem and cement during and after polymerisation in the presence of firm peripheral anchorage of the cement to the endosteal bone of the femur.

A study of the mechanical properties of the fibrous membrane found at the implant—bone interface of loose total knee arthroplasties in dogs was published 8 years ago (Hori and Lewis 1982). The authors showed that this membrane exuded fluid under compressive loads that was later reabsorbed as the loads were removed. The membrane tissue was shown to have a low elastic modulus, especially under low compressive loads.

Thus, in the presence of a defect in the cement mantle, there is a potential transport mechanism for the passage of fluid and debris from the joint cavity, down between the stem

Case 1

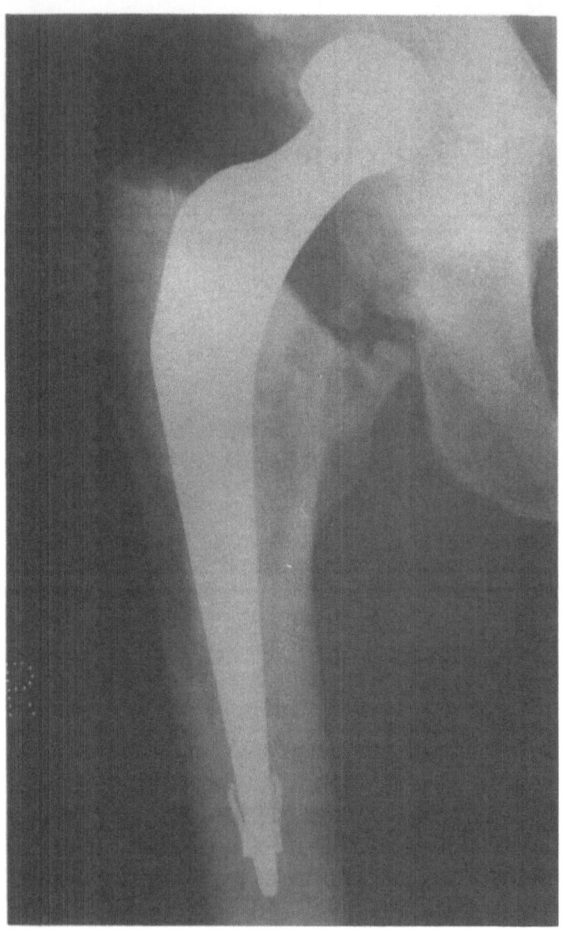

Fig. 16.1. Case 1: AP X-ray at 3½ years following hip arthroplasty. Note area of lysis in zone 2.

R.M.: farmer.
Right hip arthroplasty 20.8.79. Full recovery and return to normal farm activities. Three years later, intermittent aching in the right thigh with vigorous activity. An X-ray showed an area of endosteal bone lysis antero-laterally (Fig. 16.1). The lesion became larger and when explored on 17.1.84, a granuloma was found (Fig. 16.2). The metal surface of the stem was clearly visible through a large defect in the cement mantle.

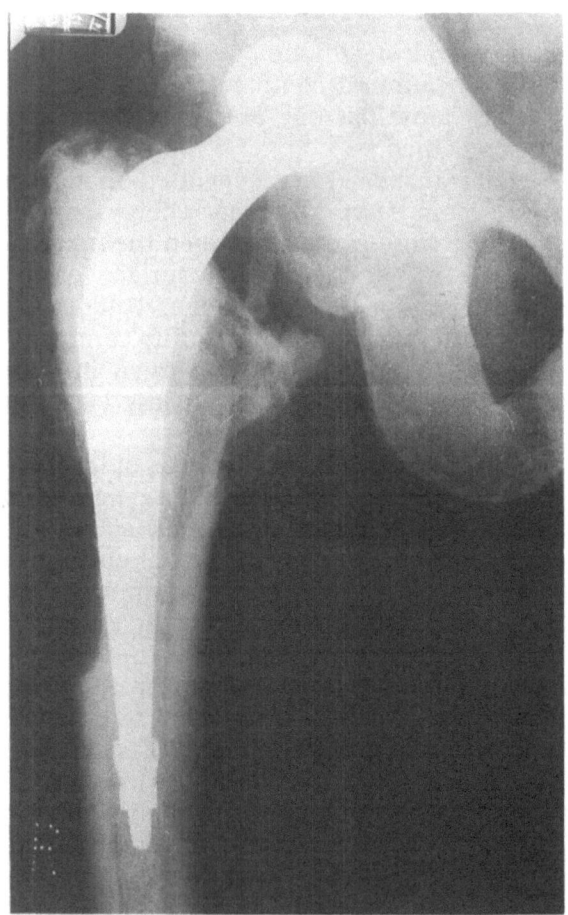

Fig. 16.2. Case 1: AP X-ray at 6 years following hip arthroplasty. Note marked increase in size of lytic area.

and the cement and out through the mantle defect onto the endosteal surface of the femur. Moreover, pressure changes occurring in the joint and at the interfaces under the loading produced by the activities of daily living may also be transmitted, with little attenuation, via the same route (Downey et al. 1988).

The histological appearance in the lytic areas of the following three cases showed granulomata in each of which was found polythene debris, fine acrylic cement debris and metallic debris. It is thought that this debris obtained access to the endosteal surface of the femur through the route already outlined. The proximal cement—bone interface was intact in each case.

Case 2

R.C.: retired.
Left hip arthroplasty 25.11.80. Post-operative X-ray showed a thin cement mantle in zone 2 (Fig. 16.3). Returned to full activities, including moorland walking. Dislocated left hip on 17.5.87, and the dislocation subsequently became recurrent. Acetabular augmentation was only temporarily successful. An X-ray on 16.5.88 showed localised endosteal bone lysis between zones 1 and 2 (Fig. 16.4). At operation, the lytic area was found to be a granuloma associated with a small defect in the cement mantle, through which the metal surface of the stem was visible in direct communication with the cavity of the granuloma.

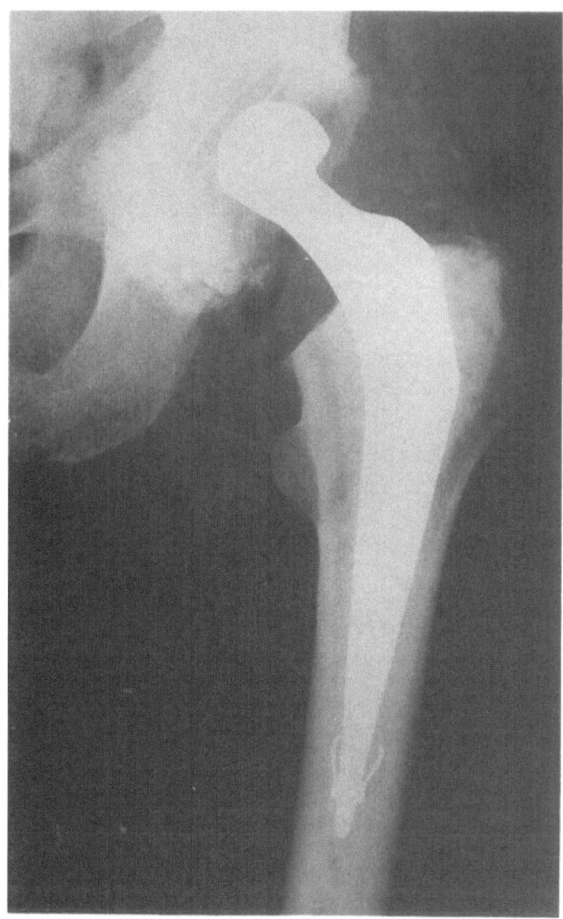

Fig. 16.4. Case 2: AP X-ray 7$^{1}/_{2}$ years following operation. Note localised endosteal bone lysis in zone 2.

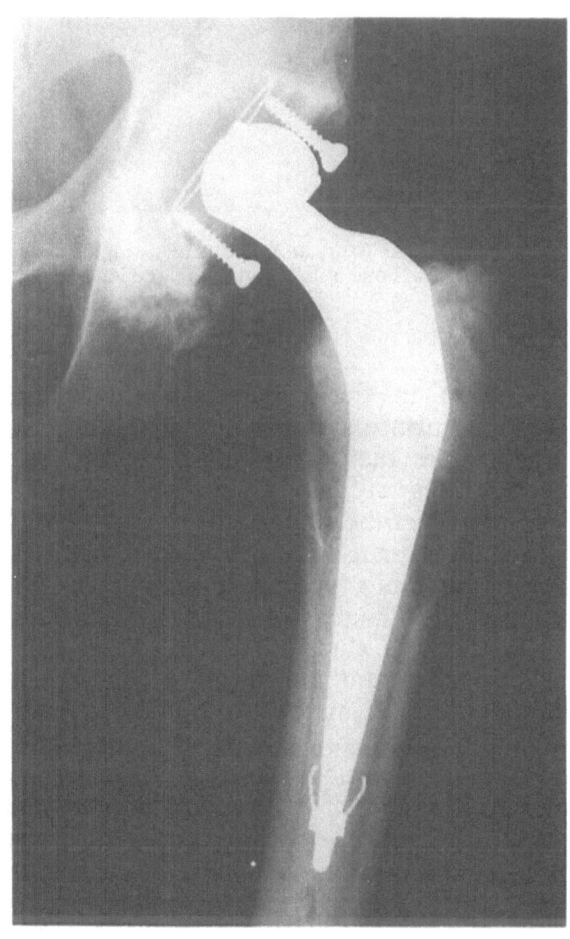

Fig. 16.3. Case 2: Post-operative AP X-ray.

Case 3

G.W.: bar manager.
Left hip arthroplasty 30.4.85. Oversized stem put into slight varus with consequent deficient cement mantle in the lower part of zone 7 (Fig. 16.5). Full functional recovery. Aching in the thigh 3 years later. An X-ray showed endosteal lysis opposite the medial side of the middle third of the femoral component in association with the cement mantle deficiency (Fig. 16.6).

Arthrography using low-viscosity dye showed the dye in the lytic lesion within 40 minutes of injection. At operation, methylene blue dye was first injected into the hip joint cavity. The lytic lesion was then explored

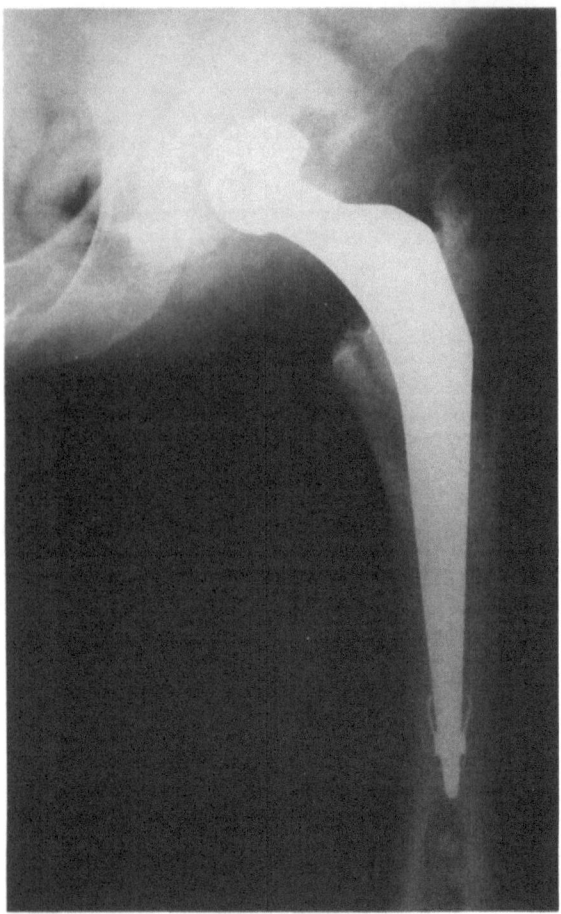

Fig. 16.5. Case 3: Post-operative AP X-ray. Note slight varus position of stem and deficiency of cement in the lower part of zone 7.

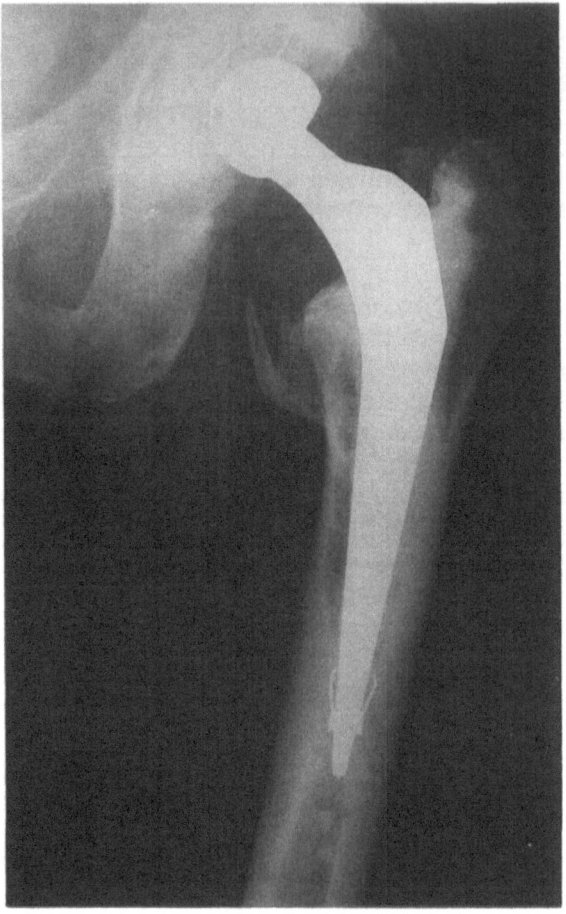

Fig. 16.6. Case 3: AP X-ray at 3½ years following operation showing endosteal bone lysis in zone 6 and the lower part of zone 7.

directly, before the hip joint cavity was opened. A defect in the cement mantle was found, through which blue dye had passed to stain the whole of the cavity of the granuloma. The hip joint was then opened and the stem removed. The blue staining was found between the stem and the cement. The proximal cement—bone interface was intact and was not stained by the dye.

Conclusions

The third patient in particular provided very strong evidence that there was a potential pathway from the joint cavity, down between the stem and the cement and out through any defect in the cement mantle to the endosteal surface of the femur. With the pressure changes that occur in association with the activities of daily living the endosteal bone may then be subjected to movement of fluid that contains any debris that might be present including polythene, acrylic cement and metal. A source of acrylic cement and metal debris is abrasion between the matt-surfaced stem and the inside of the cement mantle (Hales et al. 1990). Polythene debris can only come from within the joint cavity itself. Bone destruction and granuloma formation then occur as consequences of the combined effects of the debris, pressure changes and fluid movement. This represents one mechanism for the production of localised endosteal bone lysis in the femur following cemented total hip arthroplasty.

Acknowledgement. The material in this paper forms part of a more extensive paper on the same subject already accepted for publication by the Journal of Bone and Joint Surgery (British Volume). The figures are reproduced here with the kind permission of the Editor of the British Volume of the Journal of Bone and Joint Surgery, to whom we offer our thanks.

Mr Ling, who presented this paper, is indebted to his co-authors for their collaborative work.

References and Further Reading

Carlsson AS, Gentz C-F, Linder L (1983) Localised bone resorption in the femur in mechanical failure of cemented total hip arthroplasties. Acta Orthop Scand 54:396—402

Downey DJ, Simkin PA, Taggart R (1988) The effect of compressive loading on intra-osseous pressure in the femoral head in vitro. J Bone Joint Surg (Am) 70:871—877

Fornasier VL, Cameron HU (1976) The femoral stem—cement interface in total hip replacement. Clin Orthop 116:248—252

Hales D, Hooper RM, Lee AJC, Ling RSM (1990) The production of acrylic cement and metal debris by the femoral component in cemented total hip arthroplasty. Paper presented at the Spring Meeting of the British Orthopaedic Association, Glasgow

Hori RY, Lewis JL (1982) Mechanical properties of the fibrous tissue found at the bone—cement interface following total joint replacement. J Biomed Mater Res 16:911—927

Huddleston HD (1988) Femoral lysis after cemented hip arthroplasty. J Arthroplasty 3:285—297

Jasty MJ, Floyd WE, Schiller AL, Goldring SR, Harris WH (1986) Localised oteolysis in stable, non-septic total hip replacement. J Bone Joint Surg (Am) 68:912—919

Lombardi AV, Mallory TH, Vaughn BK, Drouillard P (1989) Aseptic loosening in total hip arthroplasty secondary to osteolysis induced by wear debris from titanium alloy modular femoral heads. J Bone Joint Surg (Am) 71:1337—1342

Chapter 17

Micro-motion: Consequences of Material Choices, Implant Design and Fit

I.C.C. Clarke

The cemented hip joint replacement (THR) has been a great boon for disorders of the hip joint. The Charnley THR is recognised worldwide as the "gold" standard of excellence in this regard, with over 25 years' clinical use. Equally successful results are claimed country by country for similar concepts, e.g. the Exeter THR in Britain and the T-28 in the United States. Since nothing is forever, well may we ask how long can this success last, particularly in the younger patients?

Cemented Implants

Clinical Results

Long-term results over 10 years have indicated that the majority of failures involved the femoral component (Table 17.1), despite the opinion in the community that the cup is more at risk (Agins et al. 1988a). In fact the

Table 17.1. Cemented Charnley 15-year survivorship (McCoy et al. 1988)

"91% survival of the arthroplasty at 15 years ... all failures involved the femoral component."

"The socket alone ... demonstrated 96% survival at 15 years."

ratio of stem revisions to cup revisions can be anywhere between 1.4 : 1 and 60 : 1, depending on which series is being reviewed. Thus the cup may look bad radiographically, but it is the stem which appears to be clinically bad, causing pain.

Polyethylene—UHMWPE Debris

Revell, Freeman and many other authors have documented the extensive bony destruction that can be brought about by the release of polymeric debris such as Teflon and polyethylene (UHMWPE) (Charnley 1961; Revell et al. 1978; Linder et al. 1983; Linder and Carlsson 1986). Although there is no doubt that bone cement can cause osteolysis (cement disease) (Jones and Hungerford 1987), the spectrum of osteolytic changes that UHMWPE debris can create by itself is awesome (polyethylene disease). The small particles become phagocytosed by macrophages which, in time, become activated invoking a series of osteolytic events in the surrounding tissues (Charnley and Kamangar 1969; Adams and Hamilton 1984).

In my experience, the majority of surgeons think that wear is not a problem, that somehow it does not happen, or if it does, it is of no consequence. This myth is probably reinforced every time the surgeon revises a total hip replacement. The implant may be painful, loose and sitting in "bad" bone stock, but at the same time the polyethylene socket looks nicely polished and the metal femoral head generally looks unscuffed. Thus the surgeon cannot be faulted for saying he sees nothing he can describe as "wear". However, one of the features of the plastic, confirmed by many laboratory studies, is that the UHMWPE wears very finely; indeed the "polished" surface is the worn surface of the plastic! Moreover, the UHMWPE particles are invisible to the naked eye and will not be detectable in routine histological examinations.

Consider then the 15-year results of the Charnley hip from the Hospital for Special Surgery in New York (Table 17.2) (McCoy et al. 1988). The average age at follow-up was 75 years, yet the majority demonstrated cup wear on radiographic follow-up. The average wear in males was 2.5 times more than in females. Polyethylene wear was found to correlate with proximal bone loss of the femur. Linear wear rates of 0.08 to 0.21 mm per year may not sound very alarming to the clinician, but from our histological studies we know that while the UHMWPE wear particles are usually around 5 μm, they can also occasionally be 50 μm. That begins to suggest very large numbers of these microscopic UHMWPE particles may account for the radiographic wear rates.

Table 17.2. Thought-provoking implant-failure concept (McCoy et al. 1988)

Average age at surgery	60 years
Average age at 15-year follow-up	75 years
Average wear, females	0.08 mm/year
Average wear, males	0.21 mm/year
"Polyethylene wear was associated with male sex."	
"Polyethylene wear was positively correlated with calcar resorption."	

If we assume all UHMWPE particles to be one size, the number of particles corresponding to the observed wear rates will rise as the particle size drops (Fig. 17.1). Using the published wear rates, we can predict that the debris being shed will number from millions up to billions of particles per year.

Assuming all UHMWPE debris was 50 μm size, the number of particles predicted would be 30 million/year. For 5 μm size, the number

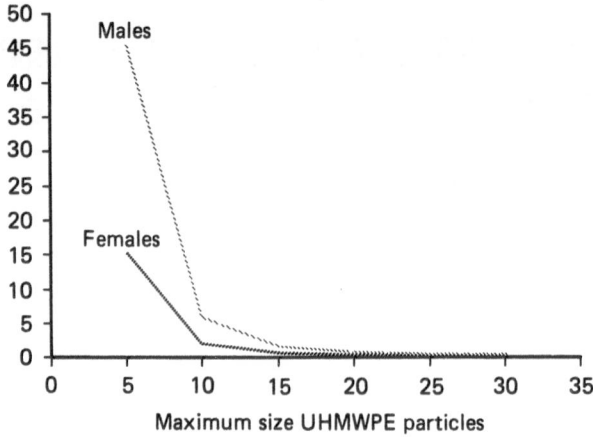

Number of particles (bn)

Fig. 17.1. UHMWPE wear particles shed per year based on 0.1—0.3 mm/year wear rates (McCoy et al. 1988).

of particles would rise to 45 billion/year, equivalent to 123 million/day (Table 17.3). How then will the tissues react to this blizzard of UHMWPE particles being shed as the patient walks on the total hip replacement? It certainly no longer appears to be something we can ignore or discount.

Table 17.3. UHMWPE wear debris

Charnley wear rates: 0.1 — 0.6 mm/year (McCoy 1988)
5 μm long particles:
  9 000 000 000 — 45 000 000 000/year
50 μm long particles:
  30 000 000 — 20 000 000/year

Survivorship curves of cemented total hip replacements in Germany show good results in the first 5 years, some deterioration in the second 5-year period and an alarming drop-off in the third 5-year period (Fig. 17.2) (Buchholz and Heinert 1988). Why is this? It cannot be overstressing, because the patients are now 10—15 years older. Perhaps it could be understressing, i.e. stress-shielding?

The results of the Charnley hips were also very good for the first 10 years at the Hospital for Special Surgery in America, but they showed a dramatic drop-off in quality during the 10—14-year period (Fig. 17.3) (Agins et al. 1988b).

So while the cemented total hip replacements are generally excellent up to 10 years, there is a common dramatic deterioration in

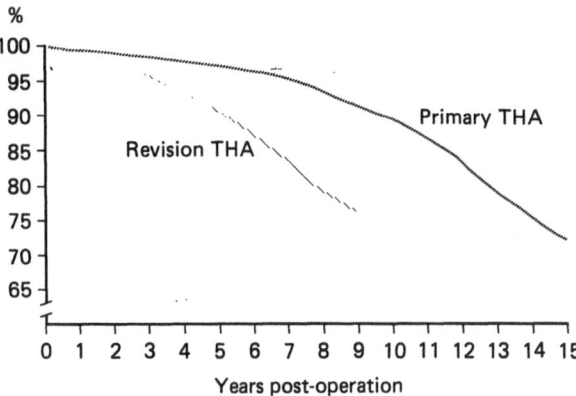

Fig. 17.2. Aseptic loosening after total hip arthroplasty. Comparison of primary and revision total hip replacements for osteoarthritis of the hip (Buchholz and Heinert 1988).

Table 17.4. Clinical "comfort" zones

Category	Years
Short	0 — 3
Intermediate	3 — 10
Long	10 — 20
Extra-long	20 — 30

the 10—15-year period. Duration of follow-up is, therefore, a very important parameter (Table 17.4). From our point of view, up to 3 years is a short-term learning curve, 3—10 years is the intermediate honeymoon period when results appear excellent but the long-term period beyond 10 years is where deterioration really starts to occur. Is it over-stressing or stress-shielding in these old patients, or an osteolytic trend dictated by the release of billions of UHMWPE wear particles?

Several authors have described the quality of hip replacement in which pain, walking and functional ability appear particularly rele-

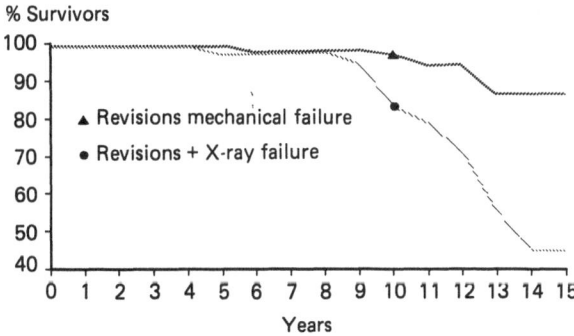

Fig. 17.3. Survivorship of the bilateral Charnley total hip replacement (Agins et al. 1988b).

vant (Charnley 1979; Hierton et al. 1983; McCoy et al. 1988). The younger patients with no walking aids and no pain have an excellent result, but they lose the most bone stock. Conversely, the least bone loss is found in patients with rheumatoid disease with pain, or other built-in limitations.

This suggests that the better the function, the faster the patients' bone stock will be destroyed, which, in turn, tells us this is an osteolytic phenomenon, driven by the release of particulate wear debris.

More than 20 years ago, the late Sir John Charnley had already detected that the rate of UHMWPE wear was twice as high in the more active, younger patients (Table 17.5) (Charnley and Kamangar 1969; Charnley and Halley 1975). Eftekhar published data of a young juvenile rheumatoid patient who wore through the Charnley cup in less than 8 years, a linear wear rate of over 1 mm per year. Once again, the UHMWPE debris can be blamed for the higher failure rates evident in the younger patients.

Table 17.5. Activity versus wear (Charnley and Kamangar 1969; Charnley and Halley 1975)

"The rate of wear is 50% greater in the active group than in the disabled group."

"The rate of wear of the under 30s group was nearly double the average for the 9 — 10 year group."

Metallic Debris

We have looked at stainless steel, Co—Cr and Ti-6-4 femoral heads wearing against UHMW-PE under laboratory conditions. One of the findings was that when Ti-6-4 heads were challenged with acrylic particles, the resulting abrasive wear could cause considerable damage to the metal head, and the Ti-6-4 metal wear particles turned the serum lubricant black. In our opinion, this was interesting but did not relate to the clinical reality. Sarmiento's series of Ti-6-4 STH hips has produced some revisions but these had no evidence of black staining (Sarmiento et al. 1979; Sarmiento and Gruen 1985). However, last year at the Hospital for Special Surgery, seven cases with a different Ti-6-4 total hip replacement were presented by Agins, in

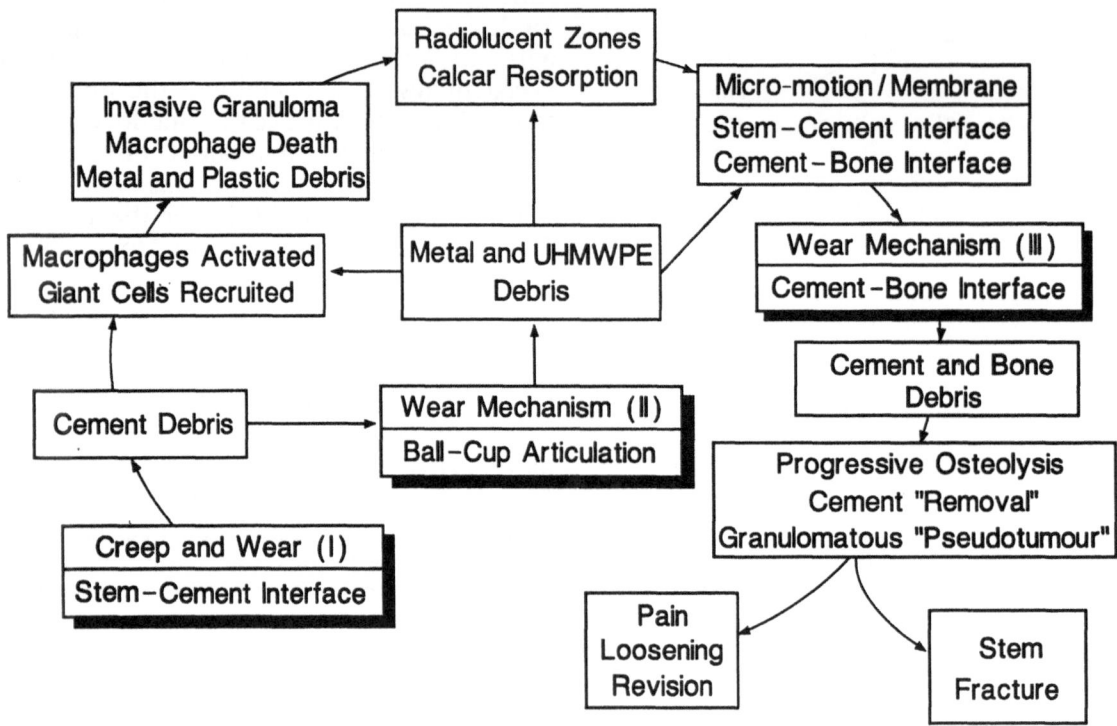

Fig. 17.4. Osteolysis model of total hip replacement wear mechanisms.

which the Ti-6-4 heads had resulted in dramatic black tissue staining, loosening and revision.

Sarmiento has now published the STH survivorship data comparing Co—Cr Charnley to Ti-6-4 STH total hip replacements. While the results appear clinically compatible there seems to be little doubt that the total hip replacements with the Ti-6-4 heads show a more rapid deterioration at the important cement—bone femoral interfaces. We consider this may be the consequence of Ti-6-4 head scuffing with more rapid abrasive wear of the UHMWPE cup, the resulting debris creating more osteolytic changes at the host interface.

Other histological studies have identified Co—Cr particles in the tissues with Co—Cr implants, and stainless steel particles when Charnley total hip replacements are used. Thus there seems little doubt that metals wear in the same way as the plastics (UHMWPE and bone cement).

In our opinion, release of wear particles is the single most important mechanism triggering osteolytic changes around the joint that will eventually and insidiously destroy

implant fixation. Thus we must focus on ways to reduce or eliminate wear debris. Any gross motion or micro-motion will set up wear mechanisms whether occurring between stem and cement, or between cement and bone (Fig. 17.4).

If the cement sheath creeps or cracks, there will be stem pistoning resulting in formation of metal and cement debris which will be ejected through any defects in the cement sheath. It may also be pumped into the joint space where it will exacerbate abrasive wear of the ball—cup articulation. The resulting metal, cement and UHMWPE debris will then be pumped around the joint, the end result being more activated macrophages and osteolysis.

Micro-motion at the cement—bone interface is disastrous for the patient, since this interface is the all-important connection to the host bone. The initial loosening may be attributed to a poor choice of materials, inadequate implant design, bad bone stock, poor bone preparation, poor cement technique or osteolysis. This bone and cement debris which is then released further insults the tissues. The end result is more activated macrophages and osteolysis, more micro-motion, pain and revision.

There appear overall to be at least three sites where micro-motion can create particulate wear debris: the cement—bone interface, the stem—cement interface and the ball—cup articulation. The first is the only biological site, the other two are mechanical involving only man-made parts with no pain receptors. Any of these three wear mechanisms have the potential to destroy the delicate balance of implant—bone fixation. Thus a spectrum of wear debris and intrafemoral osteolysis is not only predictable but a certainty, depending on where the weakest link is in the chain.

Non-cemented Implants

Metal Debris

In terms of particulates, surely the next level of escalation is to substitute metal particles instead of cement particles. Metal slurries are far more destructive than plastic debris.

There are now reports in the literature of non-cemented Co—Cr implants failing 3 months to 3 years after surgery with metal beads observed radiographically coming off the implant (Buchert et al. 1986; Rosenqvist et al. 1986). The situation is even worse now if the actual beads become intra-articular and degrade the metal—polyethylene articulations (Table 17.6).

Table 17.6. Loose beads in 22 of 34 PCA knees (Rosenqvist et al 1986)

"The loosening of beads occurred later (after 3 months) and was invariably associated with the presence of radiolucent zones."

"Intra-articular beads were visible on the radiographs of six knees."

LeTournel presented intermediate-term results with three designs of bead-coated Co—Cr femoral stems, the Lord, the Jedet and his own design (LeTournel 1988). With each stem design, there was a consistent description of beads coming off, black tissue staining, osteolytic membranes, granulomatous pseudotumours, pain and revision.

The bone destruction and progress to failure seem remarkably akin to those attributed to "cement disease" yet these cases were uncemented implants (Jones and Hungerford 1987).

We have also analysed failed Ti-6-4 bone-ingrowth surface replacements of the hip which had obviously had neither femoral stem nor bone cement to complicate matters. The articulating surfaces of the metal shells were extensively scored and the peri-articular tissues had intensive black staining. However, there was extensive bone ingrowth into the titanium fibre mesh inside the shell, so micro-motion between shell and underlying bone could be eliminated. Despite this implant stability, the femoral heads were full of giant cysts containing macrophages, titanium and UHMWPE debris. It is surprising that so much debris could be pumped into the femoral head and there still be remarkable stability from bone ingrowth. The osteolytic reaction to particulate debris was so strong, however, that it negated any benefits from bone ingrowth. Reflecting on our cemented total hip replacement wear—osteolysis—fixation—failure model (Fig. 17.4) we can simply change the "cemented" interfaces to "porous ingrowth" interfaces and the model still stands.

Instead of cement particulates acting as wear agents, there is now the real eventuality of metal debris, beads and wires circulating in the joint (Fig. 17.5). The risk is that such metal particulates will hasten the onset of osteolytic destruction and fixation failure more quickly in the non-cemented implants than in the cemented implants they replaced. In our haste to abandon cement, every problem was attributed to it under the label of "cement disease". However, it was never cement disease, it was truly a "particulate disease".

Summary

Three areas where wear mechanisms can occur have been identified to explain the spectrum of the implant loosening phenomenon; at the stem—cement interface, the ball—cup articulation and the cement—bone

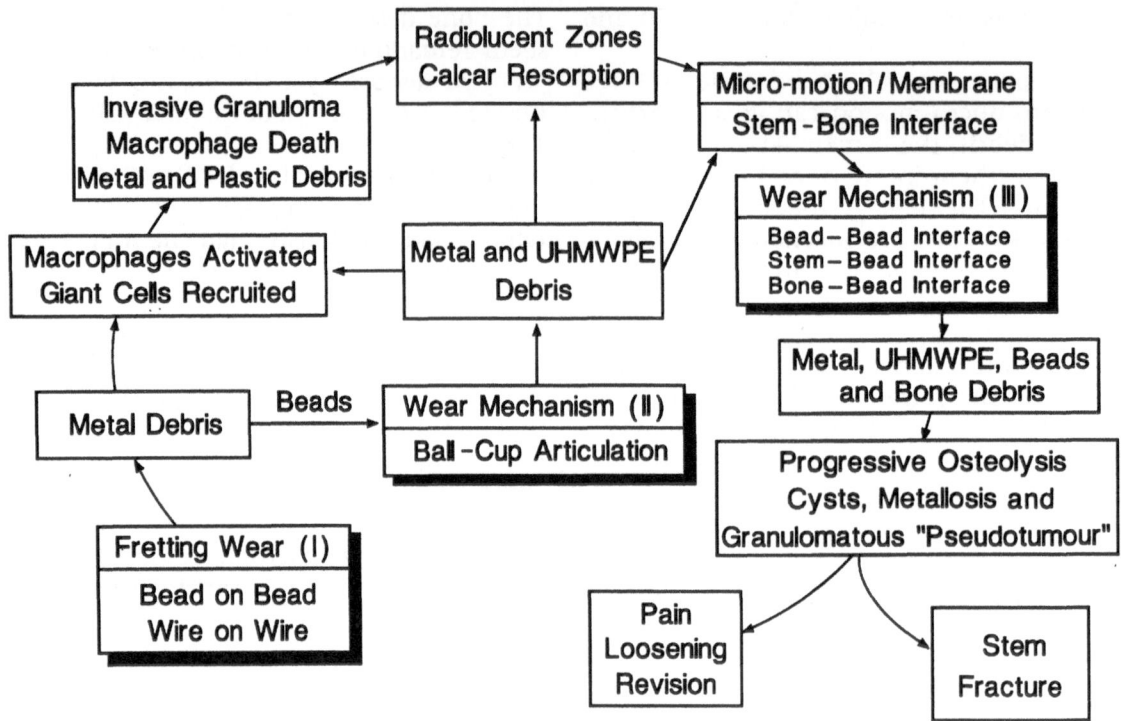

Fig. 17.5. Debris contamination.

interface. The parameters involved included the selection of implant materials, the implant designs, the adequacy of the surgical procedure, the quality of the bone stock, the activity level of the patient and the duration of use of the total joint replacement. The fate of the large man-made device is a truly a biocompatibility issue, related to the macrophage cell.

The particulate wear debris is phagocytosed by macrophages which then become activated and trigger an osteolytic response. Loss of implant fixation is seen as a progressive deterioration from the initial appearance of radiolucent zones, cysts, granulomatous pseudotumours to loosening, pain and revision. This appears consistent whether the failing implant is cemented or an uncemented porous ingrowth component. These three wear mechanisms together constitute the true Achilles heel of any implant.

The time scale also appears very important. This osteolytic process appears to take 10—15 years to progress radiographically in the older, less active patients. It may not necessarily be a clinical problem, but any upset to this equation can dramatically shorten the natural history. Many patients who live 10—15 years

after surgery can be certain of at least one revision in their lifetime.

The goals should obviously be to use the most biocompatible materials, reduce micromotion and minimise wear at the articulation of ball and cup. The easiest first stage is to abandon metal-bearing surfaces in favour of the low-friction, more wear-resistant and more biocompatible ceramic. The second stage, now well under way, is to take bone cement out of the equation. This is easy to do on the cup side but more problematical on the femoral side. The most attractive concept is a precision press-fit which has good salvage potential. A third stage must surely be to eliminate the UHMWPE, but the only way we can do that is to go back to metal—metal or ceramic—ceramic bearings. Previous experience with such combinations was not very successful.

Conclusions

There are many choices that can be made to improve the quality of the total joint replacement procedure when the basic mechanisms of the osteolytic phenomena are understood. Wear must be reduced to the lowest level possible, the interfaces must be kept clean and non-abrasive, the most compatible materials must be used and the young, active patient must be counselled.

References and Further Reading

Adams D, Hamilton T (1984) The cell biology of macrophage activation. Annu Rev Immunol 2:283—318

Agins HJ, Alcock NW, Bansal M, Salvati EA (1988a) Metallic wear in failed titanium-alloy total hip replacements; a histological and quantitative analysis. J Bone Joint Surg (Am) 70:347—356

Agins HJ, Salvati EA, Ranawat CS, Wilson Jr PD, Pellicci PM (1988b) The nine to fifteen year follow-up of one-stage bilateral total hip arthroplasty. Orthop Clin North Am 19(3):517—530

Buchert PK, Vaughn BK, Mallory TH, Engh CA, Bobyn JD (1986) Excessive metal release due to loosening and fretting of sintered particles on porous-coated hip prostheses. J Bone Joint Surg (Am) 68:606—609

Buchholz HW, Heinert K (1988) Long term results of cemented arthroplasty. Orthop Clin North Am 19:531—550

Charnley J (1961) Arthroplasty of the hip — A new operation. Lancet 1:1129—1132

Charnley J (1979) Low friction arthroplasty of the hip: theory and practice. Springer-Verlag, Berlin Heidelberg New York

Charnley J, Halley DK (1975) Rate of wear in total hip replacement. Clin Orthop 112:170—179

Charnley J, Kamangar A (1969) The optimum size of prosthetic heads in relation to the wear of plastic sockets in total replacement of the hip. Med Biol Eng 7:31—39

Hierton C, Blomgren G, Lindgren U (1983) Factors associated with calcar resorption in cemented total hip prostheses. Acta Orthop Scand 54:584—588

Jones LC, Hungerford DS (1987) Cement disease. Clin Orthop 225:191—206

LeTournel E (1988) Failures of biologically fixed devices: causes and treatment. In: Fitzgerald Jr R (ed.) Non-cemented total hip arthroplasty. Raven Press, New York, pp 318—350

Linder L, Carlsson AS (1986) The bone—cement interface in hip arthroplasty. A histologic and enzyme study of stable components. Acta Orthop Scand 57:495—500

Linder L, Lindberg L, Carlsson A (1983) Aseptic loosening of hip prostheses. A histologic and enzyme histochemical study. Clin Orthop 175:93—104

Mallory TH (1988) Mallory head/hip system. Symposium: Current concepts in implant fixation. Mount Sinai Center, Cleveland, Ohio

McCoy TH, Salvati EA, Ranawat CS, Wilson Jr PD (1988) A fifteen-year follow-up study of one hundred Charnley low-friction arthroplasties. Orthop Clin North Am 19:467—476

Pazzaglia UE, Ceciliani L, Wilkinson MJ, Cell'Orbo C (1985) Involvement of metal particles in loosening of metal—plastic total hip prostheses. Arch Orthop Trauma Surg 104:164—174

Revell PA, Weightman B, Freeman MAR, Roberts BV (1978) The production and biology of polyethylene wear debris. Arch Orthop Trauma Surg 91:167—181

Rosenqvist R, Bylander B, Knutson K, Rydholm U, Rooser B, Egund N, Lidgren L (1986) Loosening of the porous coating bicompartmental prostheses in patients with rheumatoid arthritis. J Bone Joint Surg (Am) 68:538—542

Sarmiento A, Gruen TA (1985) Radiographic analysis of the low modulus titanium femoral total hip component. Two to six years follow-up. J Bone Joint Surg (Am) 67:48—56

Sarmiento A, Turner TM, Latta LL, Tarr RR (1979) Factors contributing to lysis of the femoral neck in total hip arthroplasty. Clin Orthop 145:208—212

Sarmiento A, Natarajan V, Gruen TA, McMahon M (1988) Radiographic performance of two different total hip cemented arthroplasties. A survivorship analysis. Clin Orthop 19:505—515

Chapter 18

Differential Movement Between Implant and Bone

A.J.C. Lee

Introduction

The purpose of this paper is to examine, using basic engineering principles, what happens when an implant is inserted into bone, especially at the interface between implant and bone.

When a joint replacement implant is used clinically, it transmits load between one part of the joint and the other. If load is transmitted through the implant system, then forces must have been applied to the implant. It is a basic fact of nature that whenever a force is applied to an object, that object deflects. The nature and extent of the deflection is controlled by the magnitude, direction and type of the force, and by the appropriate stiffness of the object. In a clinical situation where a total joint implant is loaded, little can be done by the engineer or surgeon about the magnitude, direction and type of applied force. However, the stiffness (and hence, the deflection) of the device can be controlled.

The stiffness of a device may be defined as the ratio of applied force to resulting deflection. In joint replacement we are primarily concerned with bending and twisting stiffness.

Bending stiffness is equal to EI, where E is the modulus of elasticity (Young's Modulus) of the material and I is the second moment of area of the cross section. I is equal to cross-sectional breadth times depth cubed divided by twelve: $I = b.d^3/12$.

For the case of a round bar, the torsional stiffness is GJ, where G is the modulus of rigidity of the material and J is the polar second moment of area of the cross section. In this case $J = d^4/32$.

It can be seen that the material properties and the geometry of the implant are important. Both have to be considered when devices are designed, with the geometrical properties receiving particular attention. It is important to recognise that both implant system and bone will deflect in normal clinical use. The deflection of the device and the bone is modified by the musculature applied to the bone, by the geometry of the device and the bone, by the strength of the various fixations at the interfaces and by the forces acting on them. In consequence, the device may be deflected in one direction under the action of the forces upon it, while the bone is deflected in another way by the action of its attached muscles. As the device bends one way and the bone bends the other, there is likely to be relative movement at the interfaces.

If the femoral stem of a total hip replacement is cemented, for example, there are two interfaces to be considered: the interface between the metal stem and the bone cement, and the interface between the bone cement and the bone. In a total hip replacement without bone cement there is, of course, only the interface between the metal stem and the bone. In every case the vital interface is the one between the cement and the bone: this is the interface that is living.

Details of mechanical design have to be considered in all cases so that acceptable interface conditions can be achieved. The implant material which is in contact with the host material must behave in a way that is acceptable to the host. It has to be remembered that every time an implant is inserted into bone, it is inserted into dead bone. It must be the aim of the implant designer and the surgeon to create conditions that will enable the dead bone to remodel into normal living bone. To do this, it has been postulated by some authors that the stress/strain regime imposed on the bone by the implant system should be "physiological". However, no implant system yet devised will actually impart to the remaining bone loads, stresses or strains similar to those that were actually on the bone before the implant was put in place. The differences between the non-homogeneous, non-isotropic characteristics of bone and the homogeneous, isotropic characteristics of implant materials are so great as to preclude matching. Consequently, there is not at present a totally physiological implant system. Engineers and materials scientists cannot, with our present knowledge and technology, create systems and materials with the same characteristics as the bone they replace. We must, therefore, hope to create a device or system that puts acceptable stress/strain conditions onto the host bone and so encourages it to model into new bone.

The interface conditions at the bone—cement boundary will control bone remodelling and the details of the mechanical design of the implant are important in controlling these conditions. By looking at the characteristics of a number of different total hips, it is possible to determine how their mechanical design affects the conditions at the vital interface between cement and bone.

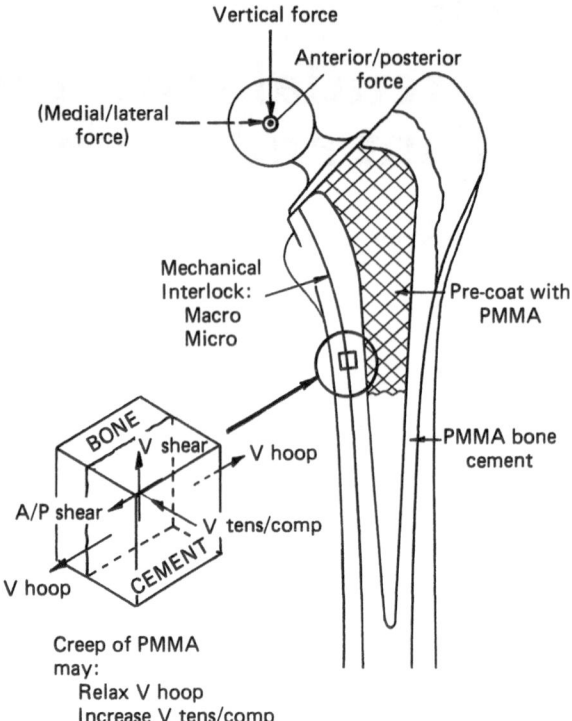

Fig. 18.1. Femoral stem pre-coated with polymethylmethacrylate bone cement.

Pre-coated Polymethylmethacrylate Bone Cement Device

This device is a femoral stem that has, at least for a portion of its length, a pre-coat of polymethylmethacrylate bone cement (Fig. 18.1). Using conventional techniques, it is implanted into bone cement. There will be vertical, medial/lateral and anterior/posterior forces acting on the head of the stem. The vertical force causes the implant to deflect downwards and to bend. The anterior/posterior force primarily causes the implant to twist in the bone, but also causes some degree of bending. The medial/lateral force primarily causes the implant to twist in the bone, but also causes some degree of bending. The medial/lateral force primarily bends the implant, but since this force is much smaller than the other two, it can be ignored, leaving only the vertical and the anterior/posterior forces to be considered.

How are these forces transmitted through the cement and into the bone? What are the resulting stress conditions at the bone—cement interface?

As a result of inserting the bone cement with proper care, there is a mechanical micro-interlock between bone cement and bone, as well as the macro-interlock caused by the tapering shape of the medullary canal. It is necessary to examine an element in the boundary between cement and bone. It should be noted that the pre-coating of the stem surface with polymethylmethacrylate will entirely lock up that interface by chemical action between the new bone cement and the coating. The strength of the interface will be considerable.

The vertical force applied to the implant will transmit substantial vertical shear through the pre-coat interface to the cement—bone interface via the cement. The effect of bending will be to put either tension or compression onto the cement—bone interface; the anterior/posterior force will also produce a shear stress at this interface.

Therefore, the primary result of inserting a pre-coated polymethylmethacrylate stem into a cemented environment is to put large shear forces on the bone—cement interface.

The time-dependent properties of bone cement have now to be considered. The effect of bone cement creep will be to relax any tensile stress present, but it will have minimal effect on shear stress. The shear stress will remain, and will be significant in the future of this type of implant.

Matt Surface Stem Device
(Fig. 18.2)

A common surface finish applied to femoral stems is the matt, or satin, finish. Measurements on a Talysurf surface roughness measuring machine reveal that the centre line average roughness of matt surface is about 1 μm (0.001 mm). The differential stiffness between the metal stem and the bone—cement combination results in some movement at the stem—cement interface. It is likely that this movement produces bone cement and metal debris which may be significant in the long term.

The vertical force on the implant will produce shear stress at the stem—cement inter-

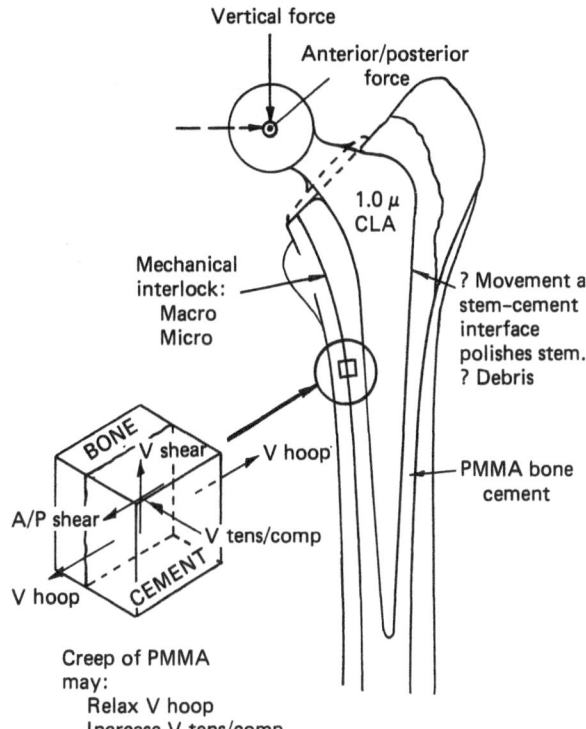

Fig. 18.2. Femoral stem with matt surface finish.

face. The friction (roughness) characteristics of the interface will allow some movement at the stem—cement interface, resulting in an increase in cement hoop stress and a diminution of shear stress. The tension/compression force due to bending will still be present, together with the anterior/posterior shear force which is controlled by the shape of the actual device.

In the longer term, following cement creep, the tensile hoop stress in the cement will relax (become smaller) leading to an increased radial compression stress in the cement. At the interface between cement and bone, the long-term stress condition will exhibit smaller vertical shear stress than with the pre-coated stem, together with similar shear stress from torsional effects and a larger direct radial compressive stress.

Polished Stem Device (Fig. 18.3)

The centre line average roughness of polished stems was measured and found to be 0.01

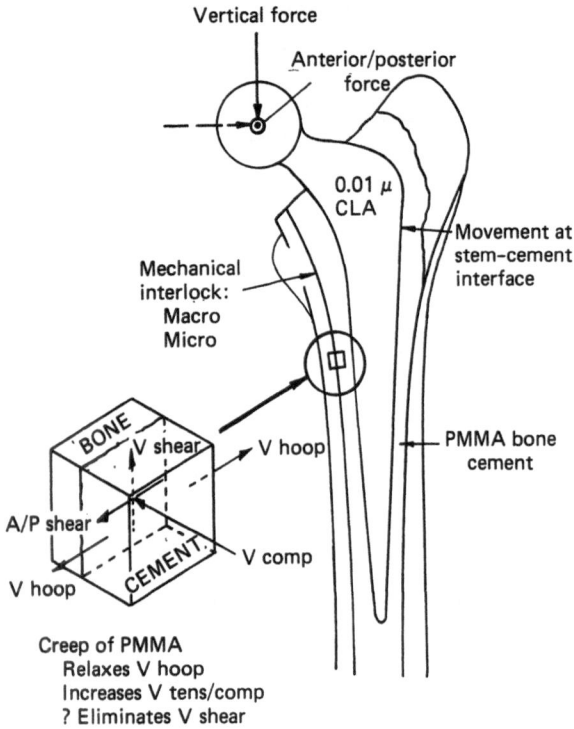

Fig. 18.3. Femoral stem with polished surface finish.

μm. Forces applied to the stem lead to the possibility of movement at the stem—cement interface (depending on the shape of the stem), but the result of the movement is different from that which occurs with a matt surface that is 100 times rougher. In particular, the production of metal and bone cement debris becomes insignificant. Very little shear force can be transmitted from stem to cement because the highly polished interface has a low co-efficient of friction. A polished stem will give substantially increased hoop stress in the bone cement, together with the tension/compression stress from bending and shear stress from torsion. In the longer term, the tensile hoop stress will relax and greatly increase the radial compressive stress. In consequence, at the cement—bone interface, vertical shear stress will be almost entirely eliminated and replaced by radial compressive stress; the shear stress arising from torsion will still be present.

Porous-Coated Cementless Stem Device (Fig. 18.4)

A porous-coated implant relies on the bone growing into the porous coating to obtain long-term fixation. Initially the implant is stabilised by mechanical interlock produced by the general shape of the implant and bony cavity (macro-interlock). Initial fixation by macro-interlock may change with time into fixation by a combination of macro- and micro-interlock, the micro-interlock being provided by ingrowth of bone or other tissue into the porous coating of the implant. There will be movement in shear at the bone—implant interface due to the different stiffness characteristics of the massive stem and the bone. Nothing can prevent the transmission of vertical shear with this system. Additionally, the interface will be subject to anterior/posterior shear and to a relatively small amount of direct compression.

The very high shear stresses occurring at the implant—bone interface are an inevitable consequence of the mechanical design of the system. Relative movement between bone and implant may be enough to "chop off" the

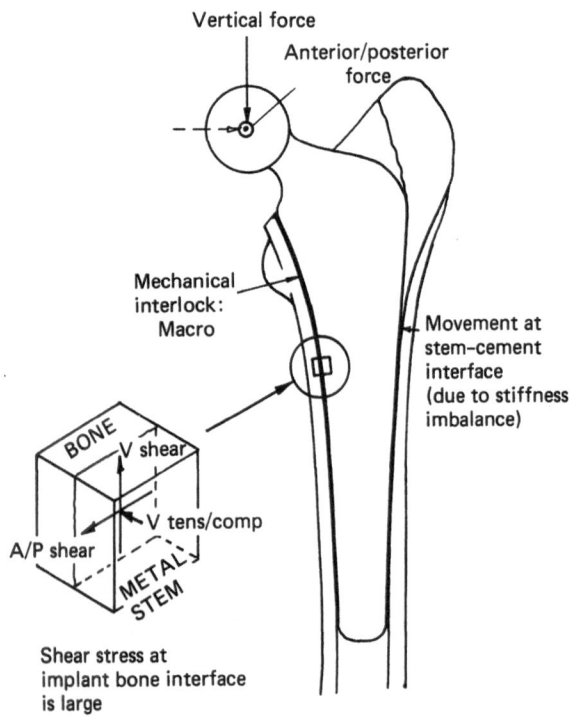

Fig. 18.4. Femoral stem, cementless, with porous surface.

bone which should be growing into the porous coat of the implant.

Conclusions

Four different types of stem have been described. A basic look at the mechanics of the four stem types has shown that the stress/strain conditions at the vital interface between implant material and bone are not the same, and must significantly depend on the mechanical design of the implant system.

Both the material and the geometry of the device must be taken into account when an implanted joint is designed. Very little can be done about the material properties of PMMA and metal but to a limited extent by changing the loading regime or by techniques of grafting for example, the shape and mechanical characteristics of the bone itself can be altered.

The aim of any implant system must be to preserve the bone stock of the joint. To do this, the interface must be loaded in a way that encourages new bone to form. From clinical observation it appears that new bone is formed when the vital interface is loaded in static compression. Implant system designers must take note of the biomechanical consequences of various types of design to ensure, as far as is possible, that harmonious designs are produced in which all parts work in a synergistic way that will lead to good long-term results.

Chapter 19

A Preliminary Report on the Stem—Cement Interface and its Influence on the Bone—Cement Interface

A.W. Miles

Introduction

This paper describes a preliminary study of how load is transmitted to the femoral bone—bone cement interface and how it is influenced by the type of stem—cement interface.

There are two diverse categories of fixation that may be used in total hip replacement. Compliant fixation is characterised by implants such as the Iso-elastic, where the low elasticity of the system allows the implant to move in concert with the surrounding bone, resulting in a more physiological loading of that bone. At the other end of the spectrum there is non-compliant fixation, characterised by a large, very stiff metal implant. This cannot move in conformance with the surrounding bone. Cemented implants fall between these two extremes, with a compliant load-mediating cement mantle interposed between the implant and bone. A considerable range exists even within cemented implants in the way load is transferred through to the cement—bone complex. There is, at one extreme, the relatively slender, polished, collarless designs such as the Exeter hip, and at the other extreme the large, stiff,

textured and sometimes pre-coated implant with a full calcar-bearing collar.

The integrity of the interfaces is influenced in the short term, reflecting the early post-operative period, and in the more protracted long term, e.g. 10 to 15 years. Short-term influences include factors such as bone necrosis associated with surgical trauma, heat of polymerisation of bone cement, cement shrinkage, curing changes and creep. In the longer term, polyethylene wear debris, acrylic fragments, cement ageing effects, bone remodelling and junctional tissue changes can all influence the quality of the interfaces (Fornasier and Cameron 1976).

Aseptic loosening is the principal long-term complication in total hip replacement and may be characterised by failure at the bone—cement interface. One suggested pattern of failure involves a mechanical phase associated with fracture of cement intrusions into bone pores, followed by a secondary phase where abrasion by particles in the interface produces a biological reaction (Johanson et al. 1987). The cycle is initiated by mechanical action at the bone—cement interface and this can be associated with a combination of the three basic types of loads — tensile, compressive and shear. Each of

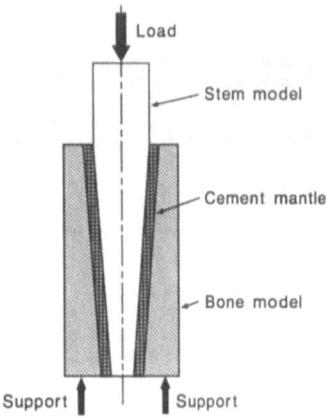

Fig. 19.1. Schematic diagram of experimental model.

Fig. 19.2. Location of embedded strain gauges in cement mantle.

these loads has related influences on the bone—cement interface. Shear is undoubtedly the most adverse and can be identified with fracture of the cement and bone, the production of fracture debris and associated abrasion in the interface leading to adverse tissue reactions. Compression, while leading to creep, would appear to be the best-tolerated load at this interface. The types of loads transferred to this interface are strongly influenced by the design of the implant and it was the aim of this study to develop a simple experimental protocol to explore some of the basic mechanisms of load transfer that occur between the stem and the cement, and the interface into which it is fixed.

Methods

An experimental model was developed consisting of a stainless steel tapered cylinder, representing the femoral stem, cemented into an aluminium sleeve with a tapered bore, representing the femur, with an interposed 4-mm thick cement mantle (Fig. 19.1). The stem taper was chosen to follow the area reduction ratio used in the Exeter femoral prosthetic stem.

A number of different surface finishes were tested. This paper presents the results of two extremes, a highly polished stem allowing minimal interfacial bonding between the stem

and cement, and a grooved sandblasted stem providing a strong macro-lock between the stem and cement. The surface finishes achieved with the different stem models were comparable with those of commercial implants (Table 19.1). The cement mantle was set at 4 mm and this was achieved by matching the taper bore of the aluminium sleeve with the taper of the stem models, and using an alignment device to insert the stem model into the cement. The bone—cement interface was standardised by incorporating shallow circumferential grooves in the taper bore of the aluminium sleeve simulating complete "osseo-integration" at the "bone—cement" interface.

Table 19.1. Surface finishes obtained for the stem models

Surface finish	Ra (model) (μm)	Ra (Exeter hip) (μm)
Polished	0.04	0.05
Sandblasted	1.50	1.20
Grooved	1-mm grooves	—

In order to record the strains within the cement mantle, strain gauges mounted on perspex carrier bases were positioned in the mid-section of the model (Miles and Dall 1985). The gauges were located 2 mm from both interfaces and arranged to measure axial and hoop strains within the cement (Fig. 19.2). The test models were loaded in axial compression through a steel ball to eliminate

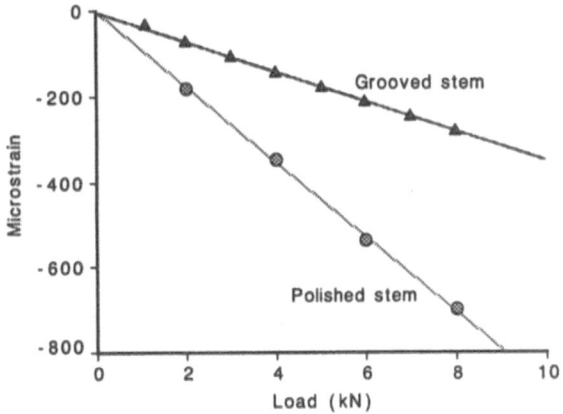

Fig. 19.3. Axial strain measurements within the cement mantle for the two stem models.

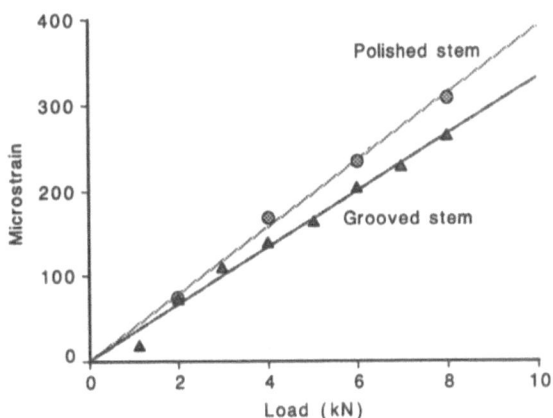

Fig. 19.4. Hoop strain measurements within the cement mantle for the two stem models.

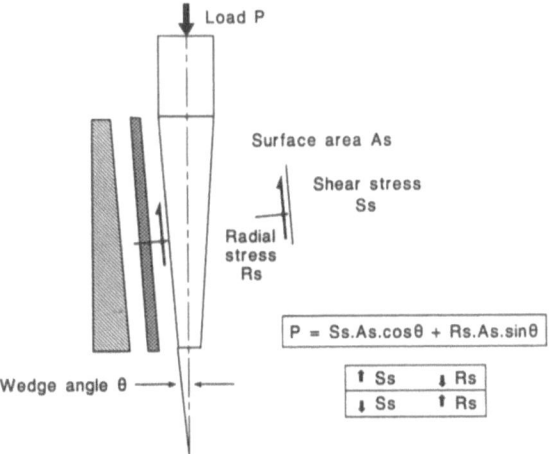

Fig. 19.5. Free body diagram of the model illustrating the balance between shear and radial stresses (Miles et al. 1989).

any bending moments. Pure axial loads of up to 8 kN (approximately 2000 lbs) were applied in a number of load increments. To ensure the stems were properly "bedded" into the cement, the complete load cycle was repeated at least 10 times and the strain readings averaged for the last five load cycles.

Results

Substantial differences were found in the axial strains for the two different surface configurations (Fig. 19.3). There was a considerably higher axial compressive strain associated with the polished stem than the grooved stem, whereas there was not much difference in the case of the measured hoop strains for the two configurations (Fig. 19.4).

These findings can be qualitatively explained by reference to a free body diagram (Fig. 19.5). The load on the femur is counteracted by a shear stress and a radial stress acting on the surface of the stem. The forces can be resolved in the axial direction to "give" at the vertical components of the load, and it is clear that by increasing the shear stress at the interface the radial stress will be reduced, and vice versa. This agrees with the experimental results. In the case of the polished stem, where there was a very low interfacial shear stress between the stem and the cement, the bulk of the load was transferred through radial compression of the cement.

The hoop strain is very dependent on the degree of radial restraint offered to the cement by the surrounding bone. This can be illustrated using a simple diagram which shows a horizontal cross section taken through the model (Fig. 19.6). When the stem is displaced vertically through a fixed amount, the influence of the degree of restraint offered by supporting structures may have clinical implications. Where the bone stock of certain categories of patients may not provide adequate structural radial restraint, it precludes the use of implant designs which induce high radial loads and associated hoop strains.

The load transferred through the cement to its interface with the bone can also be represented by a simple diagram (Fig. 19.7). In the

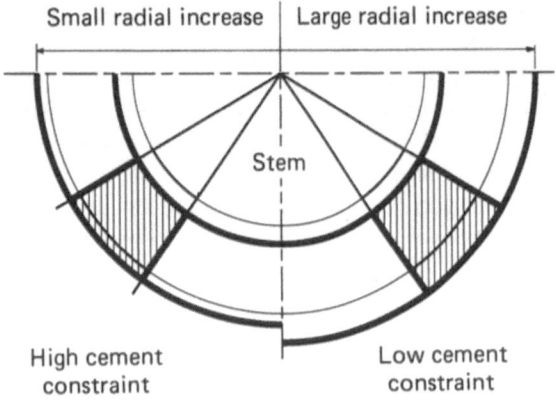

Fig. 19.6. Cross section of model illustrating the influence of the radial restraint.

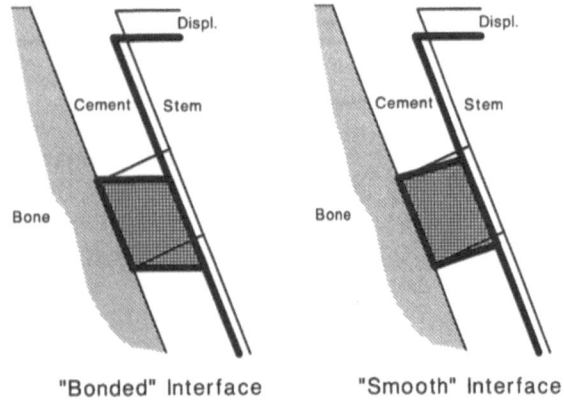

Fig. 19.7. Diagram illustrating the load transfer through a rectangular element within the cement mantle for both types of stem models.

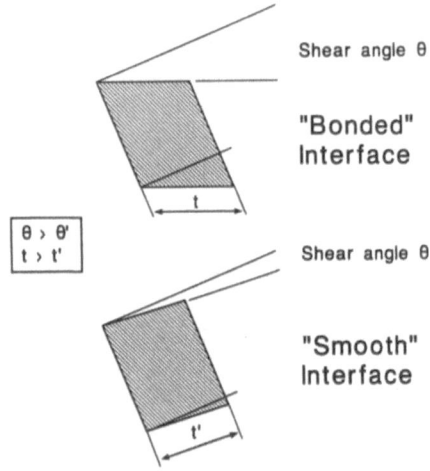

Fig. 19.8. Diagram illustrating the shear distortion of the rectangular elements of cement depicted in Fig. 19.7.

cross section representing the cement it can be seen that there is considerably more distortion in the rectangular element of cement when the bonded stem undergoes a fixed vertical displacement, whereas the smooth stem allows relative slip between the stem and cement and consequently there is less distortion.

The angle shown in Fig. 19.8 is a measure of the shear strain imposed on the element. It shows that this shear strain and hence the shear stress is much greater in the case of the bonded stem than in the smooth stem. There is, however, a greater degree of radial compression in the case of the smooth stem since the cement is compressed radially into a smaller space. These qualitative observations based on a simple diagrammatic representation concur with the experimental findings. After the completion of the strain measurements, the stem models were removed from the aluminium sleeves in reverse push-out tests. The load required to remove the smooth stem from the cement was 5.6 kN (Table 19.2). The remaining cement mantle was pristine and glassy in appearance. In contrast, a sandblasted stem took three times the load to extract, the remaining cement mantle was matt in appearance and particles of acrylic had adhered to the interstices of the stem surface. A sandblasted stem which may appear to be quite smooth does, in fact, abrade debris away from the bone—cement interface. The grooved stem took a load of more than 70 kN and could not be extracted so the model was cut in half to remove the stem.

Table 19.2. Results of reverse push-out tests on the three different stem configurations.

Surface finish	Load (kN)
Polished	5.6
Sandblasted	15.8
Grooved	>70.0

Conclusions

These preliminary tests have demonstrated that differences exist in the mode of load transmission to the bone—cement interface for the two extremes of stem surface finishes. A high degree of mechanical interlock at the interface results in greater interfacial shear stress at the bone—cement interface, reduced radial compression and a marginally lower hoop stress. The hoop strain, however, is strongly influenced by the quality of the radial restraint offered by the surrounding skeletal structure.

The initial implications of these findings are that a smooth polished stem is beneficial in reducing destructive shear loading of the bone—cement interface. Furthermore, the higher radial compressive load transferred to the bone—cement interface, combined with potential to accommodate creep in the cement, might result in a system better able to adapt to bone remodelling at this interface.

Future developments in this study will concentrate on a bone model which is more representative of the structural support offered by bone in vivo. The influence of non-circular stem cross sections and the effect of bending and torsional loads will also be addressed. Finally, it is planned to examine the important influences of the cyclic creep behaviour of the bone—cement—stem complex.

Acknowledgements. The author gratefully acknowledges the contributions of his colleague Dr S. Clift and past mechanical engineering students N. Wood and S. Wainwright for their contributions to the study of which this work is a part.

References and Further Reading

Fornasier VL, Cameron HU (1976) The femoral stem—cement interface in total hip replacement. Clin Orthop 116:248—252

Johanson NA, Bullough PG, Wilson PD, Salvati EA, Ranawat CS (1987) The microscopic anatomy of the bone—cement interface in failed total hip arthroplasties. Clin Orthop 218:123—135

Miles AW, Dall DM (1985) An experimental study of femoral cement stress in total hip replacement — influence of the structural stiffness of the femoral stem. Eng Med 14:133—135

Miles AW, Clift SE, Wainwright S, Wood N (1989) An experimental and finite element analysis of the influence of the stem—cement interface in total hip replacement. Presented at 8th European Conference on Biomaterials, Heidelberg (Full paper to be published in Advances in Biomaterials by Elsevier Science Publishers Ltd in 1990)

Discussion

The Chairman - **Professor Fitzgerald**

The Panel - **Mr Ling**
      - **Dr Lee**
      - **Mr Miles**

Professor Fitzgerald: Mr Ling, now that you think you have a handle on the problem of distal endosteal lysis, how do you propose to prevent voids in the cement about the femoral stem?

Mr Ling: A defect is an area in which there is no cement between the metal and the bone. We can prevent this by the preparation of the cavity, the proper filling of the cavity, and the accurate insertion of the stem. We hope to be able to make this foolproof by using suitable jigs. There is no doubt that the polished surface is virtually immune from the complication of endosteal lysis.

Mr Johnson: May I ask Dr Lee and particularly Mr Miles, what influence do you think the difference between your models, which involve a taper within a taper, and the extra stem which is the two-dimensional taper within a cylinder, will have on the magnitude of your tests? When assessing a pre-coated stem within the cement, will the collar that actually bears on the calcar have an effect? Will that protect the bone—cement interface?

Mr Miles: I have no vested interest in the extra stem, and believe that what we are really modelling is probably the proximal half of the stem, where there is a taper both in the femur and in the stem. I do not believe that the rectangular cross section will make a great difference. In terms of the taper it is really the change in area that is important and the way that it transfers load. There will be some differences and we hope to model them later.

I have always felt that a collar with a very stiff stem, or even more a collar with a pre-coated or pre-textured stem, is in conflict in terms of load transfer. I think the collar will bypass the bone cement mantle initially and transfer load to the bone. Results have shown that there is significant calcar resorption, and the effect of the collar on the bone is soon neglected. The collar then bears on the cement. I think that is going to make the shear at the bone interface much worse, because you are bypassing the load-mediating properties of the cement to transfer load through from the interface to the bone—cement interface.

Dr Lee: As I tried to indicate, the loading and the stress/strain conditions in the patient are extremely complicated. We have a number of clinical results and have obviously been looking very carefully at them. The way to understanding lies through developing suitable models. Quite plainly, you must first try to understand the relatively simple and then move on to the more complex and realistic systems.

Even in the simple models, though, I believe that you can model certain characteristics such as the effect of surface roughness and some differential movement. I believe that our models do not tell the whole story. By these tests we are just throwing little spotlights on some aspects of the problem. Illuminating the whole problem more and more with these shafts of light, we should eventually come to understand the more complex total system.

Professor Willert: In the cases where we observed the osteolysis, the development was years after implantation. So the bone cement defect is present right from the beginning when the prosthesis is inserted. How can one explain this long interval until the osteolysis develops and material such as polymethylmethacrylate assembles there? If this is what you assume, then micro-motion between the stem and the cement must be in existence immediately after the insertion of the prosthesis.

Mr Ling: It is a very good question. The first two cases that we saw were both in heavy, active men, and it was 7 years before they presented with this problem. We found polyethylene in both of these lesions. The third presented much earlier, at about $3^1/_2$ years.

How can debris get down between the stem and the cement? Aside from Fornasier and Cameron's findings, there is another potential mechanism. In your cases I imagine that those stems were matt surfaced. Acrylic particles will not produce much debris straight away. With time, and the effects of the anteriorly and posteriorly directed components of joint force, I suspect that creep will gradually enlarge the mantle. As that happens the movement of the stem inside the mantle will increase slightly, and therefore it will become more abraded. This exactly fits the polished appearance that we have seen on half the front and half the back of these stems. They could not do that if there was not room for them to move. The defect is probably there from the beginning. It may be very small, but if there is a passage of fluid through that defect it may actually get larger. I do not find it difficult to understand why the presentation should be late on those grounds.

Dr Clarke: I have seen 14 patients where cysts appeared 1 year after surgery which were quite alarming by the third year. Many years' previous experience by the same surgeon with the Charnley prosthesis failed to show such cysts. It is interesting that it does not take 7 years; in some designs it can be seen in a year. The stems are close to the endosteal wall posteriorly and laterally because of the design and positioning of the implant.

Dr Isaac: Could Dr Clarke possibly offer an explanation for the difference between the low wear rate found in the laboratory and the very high wear rates found in the clinical situation?

Dr Clarke: In the University of Southern California the wear rates in the laboratory correspond very well to those published by the Charnley group for Charnley hips.

Dr Isaac: No, I mean the polyethylene wear.

Dr Clarke: Contamination may occur in a very clean situation. There can be acrylic debris present or metal particles or even beads. The wear can be exacerbated in one hip joint. Even bilaterally you may see aggressive wear on one side and nothing on the other. There is a delicate balance between having reasonable wear rates and having a runaway type wear rate.

Professor Fitzgerald: Dr Clarke, you have made out quite a bad case against shedded beads. I could show you literally hundreds of X-rays of patients with PCA implants with shedded beads who are doing marvellously well probably 2 to 7 years after surgery. The bead shedding occurred early and has not progressed. It probably related to some subsidence as well as rotation on the acetabular side, and has settled down.

Dr Clarke: I would not wish to indicate that every X-ray showing beads coming off points to alarming failure rates. Looking at it overall, I would be concerned about beads coming off and exacerbating wear at the interfaces, and seeing the beads embedded in acetabular cups and in the polyethylene tibial trays. When you have metal beads articulating against femoral condyles you will wear even the cobalt—chrome ball.

Professor Fitzgerald: It is going to be difficult to get a bead inside a hip, HDP or articular surface. It can be done and certainly will occur, but it is a rare rather than common phenomenon.

Dr Clarke: In self-defence here, nothing is for ever! A Charnley may do very well for 10 years, a PCA may do very well for 10 years, but what happens after that?

Professor Fitzgerald: There is now evidence that many Charnley hips do well for over 20 years. You cannot write them off at 10 years. Certainly there is a potential problem, but it is not necessarily a lethal one.

Mr Ling: Further to what Ian Clarke has said, another mechanism by which debris and any beads can get into the articulation is through the momentary subluxation that occurs at the extreme of flexion. It is common to find patients who can flex through 130°. If you articulate the components, that range of flexion is beyond the range allowable without impingement, so the head will lift out of the socket. As soon as that happens there must be some suction effect into the socket; the head then falls again releasing the suction. A bead or debris within the socket is trapped.

Dr Clarke, it seemed to me that, reading the two studies taken from the Hospital for Special Surgery, first the 10-year study and then the 15-year study of the Charnleys, there was a huge change in the measured penetration at 15 years. The figures were in the region of 18% at 10 years and 84% exceeding 1 mm at 15 years. That suggested to me that there was some change in the behaviour of the bearing in this later period.

Professor Dowson in Leeds said that he thought this might be associated with sub-surface fatigue in the polyethylene. If that is the case, then changing the ball to ceramic will not alter the situation, because it is related to stress and not to abrasion. Have you any comments?

Dr Clarke: That is certainly one possibility. Another is the degradation of the cement. There is a worsened abrasive wear situation with polyethylene. I do not think we have any detailed knowledge on what is happening at 10 or 15 years. Looking worldwide it is alarming that the survivorship curve is falling off beyond 10, 12 and 13 years.

Professor Fitzgerald: I do not think that everybody agrees with that. John Charnley's data and Wroblewski's follow-up showed no rapid fall-off. There may have been a gradual diminution of the excellent results but it is not "falling off the cliff". What are your 15-year data, Mr Ling?

Mr Ling: I agree with what you say, especially with regard to the femoral side where we are not seeing any sudden degeneration. It may be different around the socket.

Professor Solomon: I have a question for both Mr Ling and Mr Miles. I should like to pick up this point about the erosion occurring where the metal stem comes in contact with the endosteum. Is this not something which is unique to the collarless stem? Although you say that it has occurred with the rough rather than the smooth surface, there must have been some subsidence of even the rough surface I presume. Is the erosion associated with subsidence and did you actually measure it in these cases.

If you have subsidence of a tapered titanium implant where there is a defect in the cement mantle from the outset would this not create excessive hoop stress at that point?

Mr Ling: This phenomenon, so far as we can see, is certainly not limited to collarless stems. The implication is that there is a defect in the cement mantle if metal is against bone. Femoral implants with polished stems that we used between 1970 and the end of 1975 are immune from endosteal lysis. A large proportion of those implants have subsided within the cement.

The only explanation that we can think of is that subsidence of the polished stem actually closes off the route through which access is obtained to the endosteal surface of the femur. It cannot be explained in any other way. The polished and matt surfaces are completely different.

Mr Miles: I agree with Mr Ling. In the long term fibrous tissue occurs between the stem and the cement around a metal stem with micro-movement at that interface, and the pumping effect could still occur. It is debatable whether the hoop stress with a subsiding stem would be sufficient to damage the cement if there was an inadequate mantle. It may well be dissipated quite rapidly as the stem subsides. Provided that the constraint in the cement and the bone architecture surrounding it is adequate, a small amount of subsidence will not damage the cement.

Mr Northmore-Ball: Mr Ling has said that increased rates of wear in the long term may be due to sub-surface fatigue in the high-density polyethylene. Does biodegradation of HDP exist?

Mr Ling: At the last meeting of the International Hip Society, one of the engineers from the Hospital for Special Surgery discussed degradation of polyethylene, which is due to some oxidation process. He also said that it is catalysed by one of the ions that is leeched out of titanium alloy.

Mr MacDonald: The Hospital for Special Surgery recently produced results to show that the major cause of debris in an articulating device is fracture of the polyethylene. This occurs because the stresses in either hip or knee components are in excess of the fracture stress of the polyethylene. The debris is generated by a fracture process whether it is fatigue or monotonic.

Mr Ling: Wroblewski has looked at the cups very carefully and I do not think he has seen fractures.

Dr Isaac: Our work at Wrightington on retrieved prostheses showed that the high wear rates were caused by acrylic debris on the articulating surface of the implant. The debris scratches the head of the femur causing a considerable increase in wear rate.

Dr Lee: At a recent meeting Dowson demonstrated that a 5 μm scratch on a femoral head increased the wear rate by a factor of 10 times. It does not take all that much debris to produce a 5 μm scratch. The barium sulphide in bone cement is extremely hard and will cause such a scratch. It is possible to set off a "cascade" of wear as a result of damage to the surface.

Dr Isaac: We could see scratches of 30 μm across and 10 μm deep on the femoral heads that we looked at on the scanning electron microscope. Very similar scratches can be produced in the laboratory by taking a piece of acrylic cement and rubbing it up and down on stainless steel.

Dr Clarke: Looking at wear rates around the world, they all fall into a range between 0.1 to 0.5 mm per year, except for those in a 24-year-old rheumatoid where it was 1.1 mm per year. Are you talking about the usual 0.1—0.2 range?

Dr Isaac: Of the 100 that we looked at, the mean penetration rate was 0.2 mm a year over a series of 100 patients who were all revised for loosening. The lowest could not be measured; the highest was 0.6 mm a year.

Mr Ling: Is the rate of wear linear with time?

Dr Isaac: I do not know. One never knows whether damage to the articulating surface occurred the day after the prosthesis was inserted or the day before it was removed. Looking at sequential wear statistics over a 15-year period, we have failed to find any difference in comparative wear rates at different time periods. Measuring wear rate requires standardisation of many factors.

There was a very large study by Griffith, Williams and Charnley which looked at 400 patients in which the prosthesis was giving good clinical results and no revision was anticipated. Wear rates between 0.7 mm and 0.21 mm per year were found. Although the cups had been cut in the acetabular prostheses in Malcolm's study, there appeared to be very little debris. In the uncemented Charnley press-fit cup series, there was a substantial reduction in the amount of acrylic wear debris found on the surface of the implant.

Mr Compton: May I ask the engineers about the relatively new concept of retention of the femoral neck and its effect on shear within the femoral canal, as propounded by Freeman?

Dr Lee: I think the amount of femoral neck that is retained will have very little effect on the shear and the stress conditions within the major part of the femoral canal. A U-shaped femoral neck is essential to insert a Freeman implant rather than the natural circular shape. In my opinion, there is little difference between the proximal Freeman preparation and the more conventional distal cut.

Mr Ling: I really cannot comment on it because I simply do not know. On engineering grounds, obviously, if you wanted to increase the torsional stability you could move the support either more medially or more laterally from the axis of rotation and it would essentially have the same effect.

Chapter 20

Calcar/Collar Contact in Cemented Total Hip Arthroplasty

R.H. Fitzgerald Jr

Calcar/collar contact was introduced by Harris over a decade ago, and has remained controversial among hip surgeons throughout the world. A randomised prospective study of patients with coxarthrosis without previous surgery was initiated 8 years ago. Its purpose was to study the HD2 femoral component with and without the collar to see if the laboratory data that Harris had generated were fact or fiction.

Patients and Methods

This study involved 65 patients with a mean follow-up of $4^{1}/_{2}$ years, and the data were analysed in 1988.

The term calcar contact is used in this context to mean the medial compression buttress where the neck was cut at the time of the surgery. The error of radiographic analysis was less than 3%.

The cemented sockets in the two groups (with and without the collar) were essentially the same as regards obliquity, anteversion and the bone—cement interface. The role of the acetabulum was discounted in this particular study since the acetabular component had no effect on the femoral component.

An assessment was made of alignment of the femoral component. The valgus placement was 2.1° in the collar group and 1.8° in the collarless group. It is very difficult to achieve calcar/collar contact since none of the collar devices so far designed allow the low viscosity cement to move away posteriorly from the calcar area.

Results

Maximum or optimum calcar/collar contact was obtained in half the patients with a collared device.

The pre- and post-operative hip scores were essentially the same in both groups of patients. Pain was also similar but one patient in the collarless group had severe discomfort. Three patients required revision surgery, two in the collar group and one in the collarless group. No patient with optimum calcar/collar contact has required revision surgery.

Gruen's classification was used for radiographic assessment. There were no cement fractures in patients with the collar prosthesis. In the collarless group, six patients had cement fractures in zone 7. This difference was statistically significant (p<0.05).

There was no statistical difference in the radiolucent lines one or more millimetres in thickness in zone 1. A statistically significant difference was seen, however, in zones 2 and 7 with greater frequency occurring in patients with a collarless prosthesis. There was no statistical difference in zones 3 to 6.

Femoral neck resorption was assessed. There was a statistically significant difference both on the endosteal and periosteal measurements of the calcar area with resorption occurring frequently with the collared prosthesis compared to the collarless prosthesis (p<0.05).

A subsidence of 0.5 mm occurred in the collared prosthesis compared with 2.15 mm in the collarless group. This difference was statistically significant (p<0.05). Calcar resorption was also more common with the collarless prosthesis at a statistically significant level (p<0.05). The radiographic changes were very worrying, and radiolucent lines could be seen in zones 1, 2 and 7 with a follow-up of just under 5 years.

Conclusion

The data from this study suggest that a collar is advantageous with a stiff stem, but time must be spent on surgical technique to achieve contact if a collar prosthesis is used.

This has led to the design of a prosthesis with a collar component which assures good collar contact. The collar itself is larger proportionally to the size of the stem, and the proximal cement spacer has been redesigned to direct the cement posteriorly away from the calcar area when it is in a low viscous state.

Chapter 21

Trial of Cementless Versus Cemented Total Hip Arthroplasty: Preliminary Results

D.W. Howie, G.C. Dracopoulos, M. McGee and C.M. Steele-Scott

The design and methods of analysis of a prospective randomised trial of cemented versus cementless total hip arthroplasty are presented. This study was undertaken because it is difficult to interpret the results of different types of total hip arthroplasties performed at different centres.

Confirmation that this study should continue was established by a preliminary analysis at the end of the first year of any worrying trends which had emerged.

Materials and Methods

Middle-aged patients were treated with either a collarless cemented Exeter hip arthroplasty or a collarless PCA cementless hip arthroplasty. Osteoarthritis was the diagnosis in every patient and in all cases there was no history of previous surgery to the affected hip. The patients were allocated by random numbers to each treatment group and the posterior surgical approach was used. The operation was performed in patients who were thought to have more than 10 years' life expectancy which justified their inclusion in such a study.

We examined 31 cemented and 25 cementless arthroplasties a year after surgery. Descriptive data used included UCLA activity score; weight to height ratios; incidence of other lower limb arthroplasties; limb restrictions; and ill health. Based on these data, the groups appeared comparable.

We were particularly interested to determine how well the patients did in the first year following arthroplasty and whether there were problems with early recovery. Patients were assessed pre-operatively and at 3, 6 and 12 months post-operatively. To eliminate bias as much as possible, the patients completed their own assessment before seeing the doctor. Assessment included a questionnaire which incorporated the Harris hip score and the Hospital for Special Surgery (HSS) score. The doctors then made their own assessment using the HSS score and examined the hip. Radiographs included an antero-posterior view of the hip and two oblique views of the acetabulum. An independent observer reviewed the radiographs.

Zonal analysis of the AP and lateral radiographs was used to determine the incidence of lucent lines at the bone—implant interface and various methods of measurement of migration and subsidence of the components were used. Measurement of the endosteal to outer cortical diameters on the pre-operative radiographs at two levels below the lesser trochanter demonstrated that the shape of the proximal femurs was similar in both groups.

Clinical Results

There was no obvious difference in recovery between those with the cemented arthroplasties and those with cementless arthroplasties and the progress seemed to be similar. The doctors thought that the patients were doing a little better than the patients themselves thought. Of interest was the presence of thigh pain in both groups.

Radiographic Results

There was a similar distribution of ectopic bone around the cemented and cementless arthroplasties. We found a high incidence of radiolucencies at the bone—implant interface of the acetabular components. Any radiolucency, no matter how thin, and even if it did not extend completely across one zone, was recorded. Approximately half the acetabular components in both groups had some radiolucency in any one zone. Invariably, these radiolucencies were not present immediately after surgery. The radiolucencies were usually 0.5 mm thick and we did not see any above 1 mm with a continuous radiolucency in all zones.

Around the cemented femoral components a thin incomplete radiolucency was rarely noted in the proximal zones. Radiolucencies were rarely found around the cementless components at the porous-coated zones proximally but radiolucencies were seen around the smooth distal part of the components in 25% of the cases. We did not detect any migration of the implants within the errors of measurement techniques. Occasionally loose beads were seen around the acetabular and the femoral cementless components.

Conclusions

The methods of analysis of a randomised study comparing cemented versus cementless total hip arthroplasty have been described. A preliminary analysis in the first year showed that clinical recovery appeared to be the same in both groups. Acetabular radiolucency appeared to be similar in the two groups but there were some differences around the femoral components.

This preliminary analysis was undertaken to determine whether we were justified in continuing this study. No adverse trends in either group were identified so that although the numbers of patients studied was small, it was felt that this and other randomised studies were justified.

Acknowledgement. Professor Howie, who presented this paper, is indebted to his co-authors for their collaborative work.

Chapter 22

Intramedullary Cement Fixation: A Comparison of the Fixation of Custom-Made Prostheses with the Fixation of Standard Joint Replacements

G.W. Blunn and M.E. Wait

Intramedullary cement fixation has become the most widely used method for securing both standard total hip replacements and massive prostheses.

A study was made of bone remodelling at the interface between bone and the cement around massive custom-made prostheses and then compared with the remodelling of bone at the interface around standard femoral components of total hip replacements.

Materials and Methods

We obtained postmortem specimens of 40 massive prostheses with their associated bone and 11 femoral components from total hip replacements. All were judged to be clinically stable at the time of death. The massive prostheses were custom-made and were used either for the treatment of bone tumours and their associated destructive lesions, for post-traumatic, degenerative or developmental conditions or after previous surgery. The total

hip replacements were used to treat patients with osteoarthritis or rheumatoid disease.

We investigated the bone—implant interface and tried to attribute the remodelling of bone to load-adaptive bone formation, to stress protection or to the redevelopment of the blood vessel supply to the diaphyseal cortical bone.

Effect of Inserting Intramedullary Cement and Stem

The effects of preparing the intramedullary cavity and inserting cement can result in localised bone death due to release of heat and monomer from the polymerising acrylic cement, to destruction of the intramedullary nutrient artery which supplies the inner portion of diaphyseal cortical bone, to reeming and packing that bone with acrylic cement or to the design of the prosthesis and resulting mismatch of the elastic modulus between the

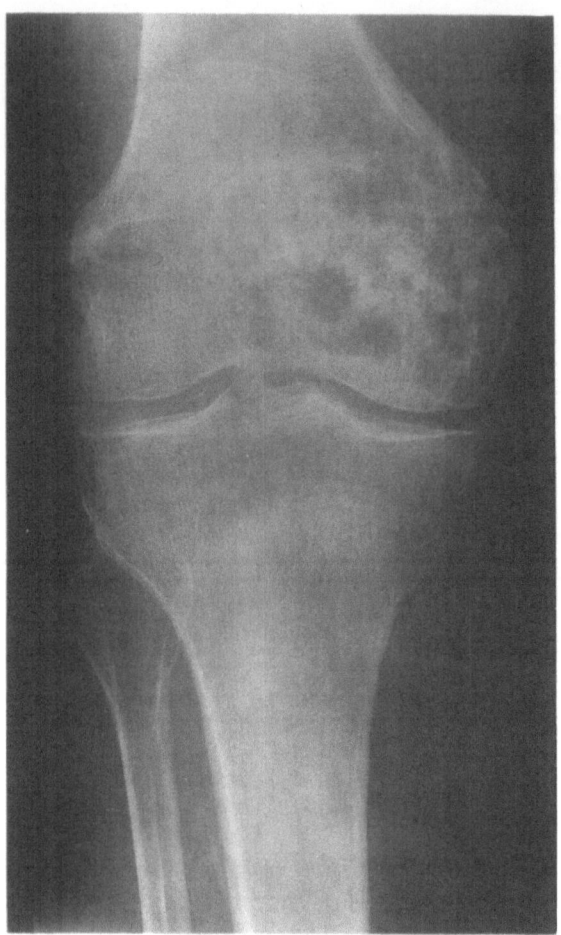

Fig. 22.1. Pre-operative radiograph showing chondrosarcoma in medial condyle of right femur.

metal—cement composite and the bone, which can alter the loading conditions of the remaining bone due to the non-physiological transfer of load from the prosthesis.

Blood Supply: Proximal Femur

The blood supply to the proximal femur is from metaphyseal vessels which supply the proximal cortical bone. An ascending nutrient artery enters mid-shaft and anastomoses with the metaphyseal vessels. Branches from this artery supply at least the inner two-thirds of diaphyseal cortical bone. The outer one-third of cortical bone is supplied in some instances by sub-periosteal vessels. Insertion of an

intramedullary stem, therefore, will destroy the blood supply to the inner diaphyseal cortex and block the regenerative path of the main artery.

Case Reports

Case 1

The patient was a 69-year-old man who developed a chondrosarcoma of the medial

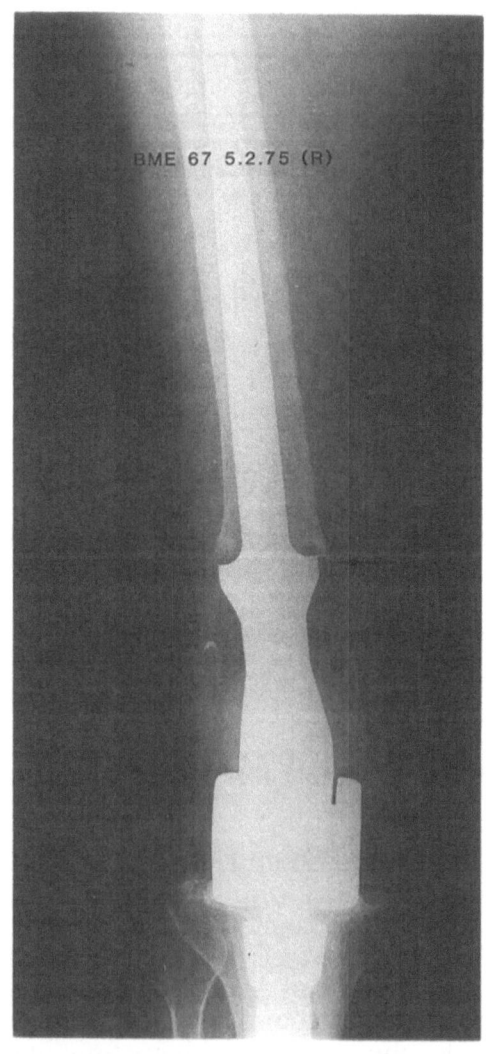

Fig. 22.2. Post-operative radiograph taken 3 years after surgery showing radiolucent space between shoulder of the prosthesis and bone. Note the small pedicle on the medial side of the femur (*arrowed*).

condyle of the knee (Fig. 22.1). The tumour was replaced with a massive distal prosthesis constructed from commercially pure titanium (Ti160). The hinged knee joint consisted of cobalt—chromium—molybdenum alloy bushes which were located in lugs of tibial components. The femoral component which carried ultra-high-molecular-weight polyethylene (UHMWPE) bushes rotated on a fixed axle.

The patient was walking unaided 6 months after surgery. Twelve years later, at the age of 81, he remained mobile but needed a stick. The following year he died from a cerebrovascular accident. There was no clinical evidence

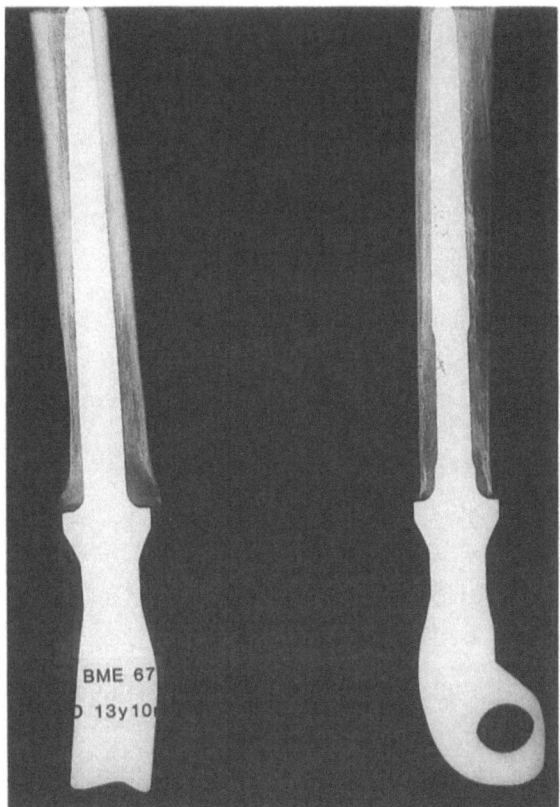

Fig. 22.4. Radiographs of retrieved prosthesis.

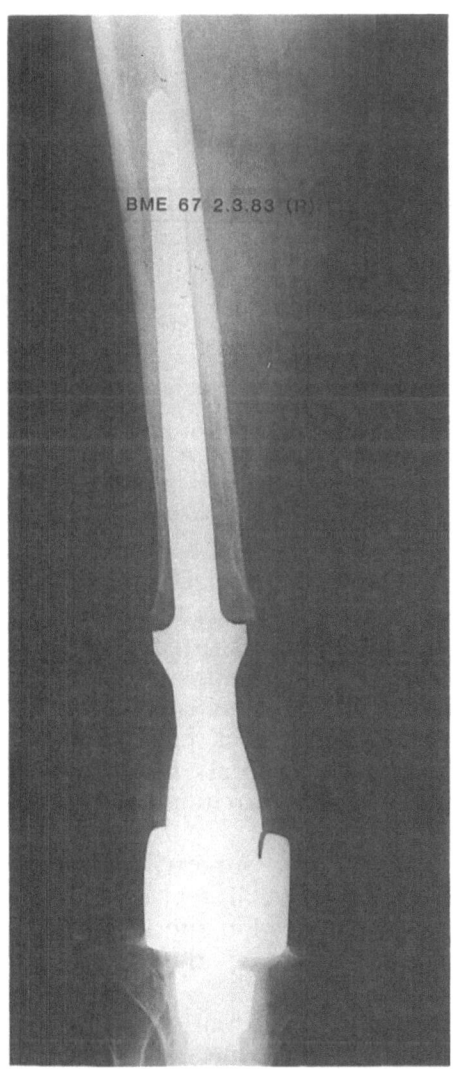

Fig. 22.3. Post-operative radiograph taken 11 years after surgery showing loss in bone density of the distal part of the femur.

of recurrence of the tumour or loosening of the prosthesis.

A radiograph taken 3 years after surgery showed a slight radiolucent space between transected bone and the shoulder of the prosthesis. A slight overgrowth of bone occurred on the medial aspect of the femur in this region (Fig. 22.2). This pedicle of bone is a common feature and usually occurs on the medial posterior aspect of the femur and develops over the metallic shaft in patients with massive prostheses. A further radiograph taken 11 years after surgery showed that the distal part of the femur had become less dense and that the pedicle had resorbed (Fig. 22.3). Radiographs taken of the retrieved femur and associated bone showed evidence of space between the transected bone and the shoulder of the prosthesis and the femoral cortex appeared to have become thinner, particularly on the posterior aspect of the femur (Fig. 22.4).

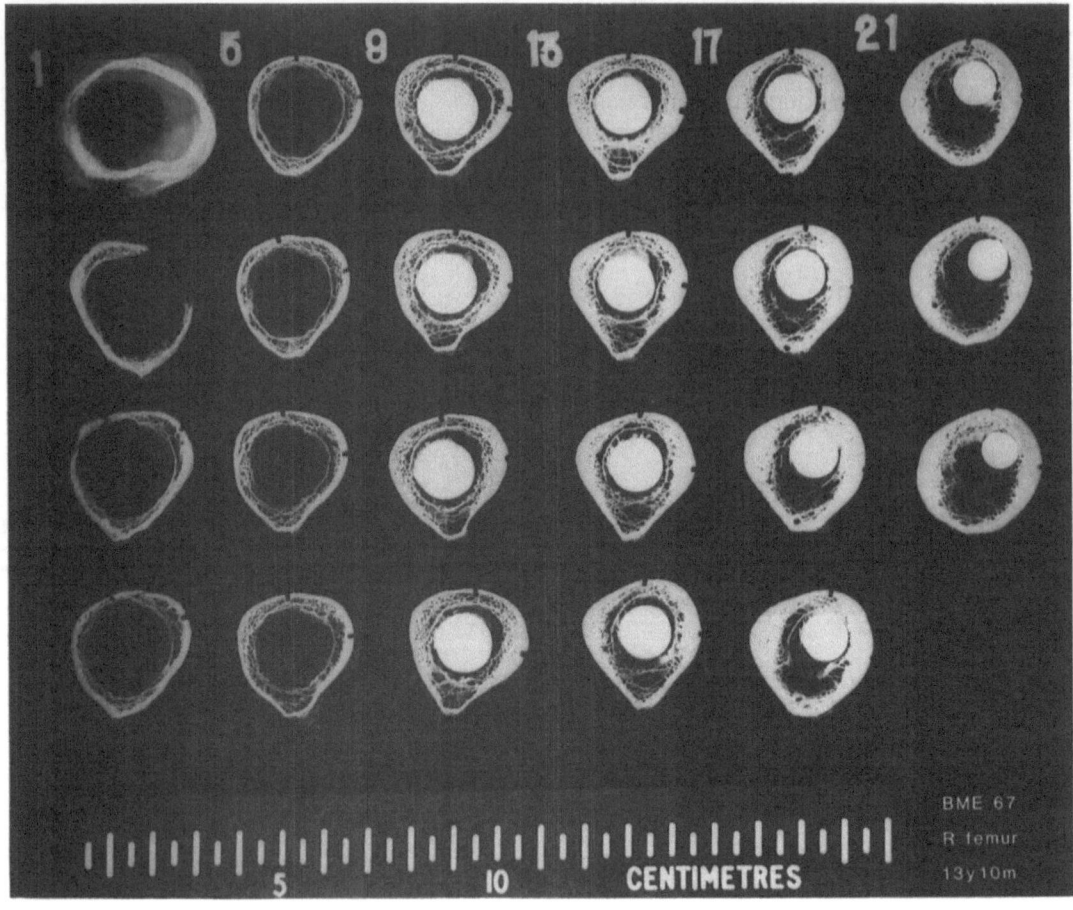

Fig. 22.5. Radiographs of sequential 5-mm thick slices with the slice adjacent to the shoulder of the prosthesis in the *top left*, through to the slice containing the stem tip (*bottom right*).

Sequential radiographs of the retrieved femur showed that the inner cortex of the femur had become porotic. A shell of bone was present, however, around the non-radio-opaque cement. This shell was strutted to the outer, more dense, cortex by trabecular bone (Fig. 22.5). Histological sections taken at the cement interface showed that a thick (20—50 µm) connective tissue interface had developed and was composed of fibroblasts and macrophages (Fig. 22.6). These cells contained small metal wear particles.

Case 2

The patient was a 70-year-old woman who developed a hypernephroma for which she had a nephrectomy. A year later a solitary metastasis was discovered in her right proxi-mal tibia. She received a Stanmore hinged knee joint, and an artificial ligament was used to attach the patella to the tibial component.

Sixteen months after surgery she suffered a sub-trochanteric pathological fracture of the right femur due to recurrence of the tumour. The proximal femur was replaced with a massive proximal femoral replacement. The patient died from carcinomatosis 2 months later.

Immediately after surgery, the radiographs of the proximal tibial replacement showed bone at the transection site in direct contact with the shoulder of the prosthesis (Fig. 22.7).

Radiographs of the retrieved tibial component and associated bone showed that a radiolucent gap had formed in this region and that a pedicle of bone had developed around the shaft of the prosthesis. Bone on the medial side of the tibia appeared less dense than bone on the lateral side (Fig. 22.8). Examination of histological sections taken through the interface adjacent to the shoulder of the prosthesis showed that soft tissue occurred next to the metal and adjacent to the acrylic cement around the intramedullary stem (Fig. 22.9a). One of the characteristics of this soft tissue was that it contained numerous blood vessels. Bone adjacent to this soft tissue appeared to have remodelled, whilst large areas of dead cortical bone, as evidenced by empty lacunae, were apparent in some areas (Fig. 22.9b).

Sequential radiographs of the retrieved femur enabled a comparison of the bone remodelling around the proximal and distal stems which had been inserted for 2 months and 18 months respectively. The cortical bone

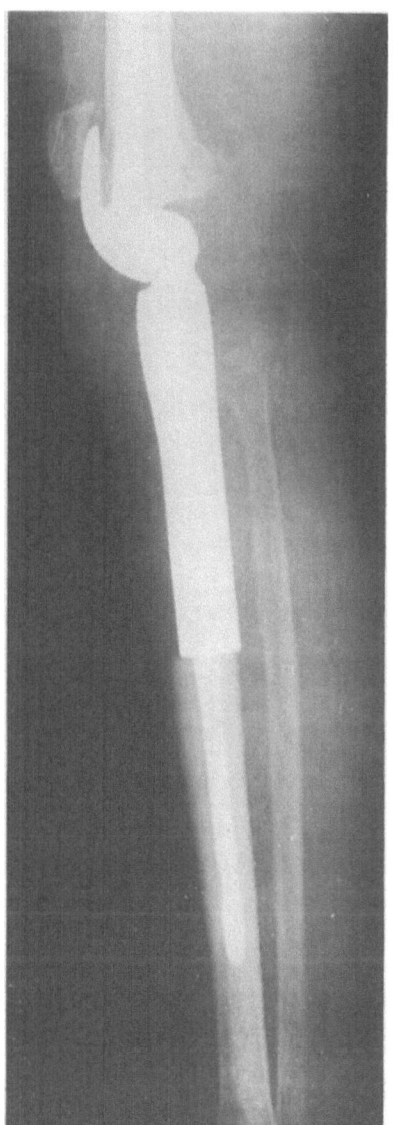

Fig. 22.7. Post-operative radiograph showing the proximal tibial component and knee joint at 1 month. Note the direct contact between bone and the shoulder of the prosthesis.

around the proximal stem remained thick and dense whilst inner cortical remodelling was evident around the distal stem. A shell of bone surrounded the acrylic cement and this was strutted to the outer dense cortex by trabecular bone (Fig. 22.10).

Case 3

The patient was a 64-year-old man who had a malignant histiocytoma in the mid-shaft of

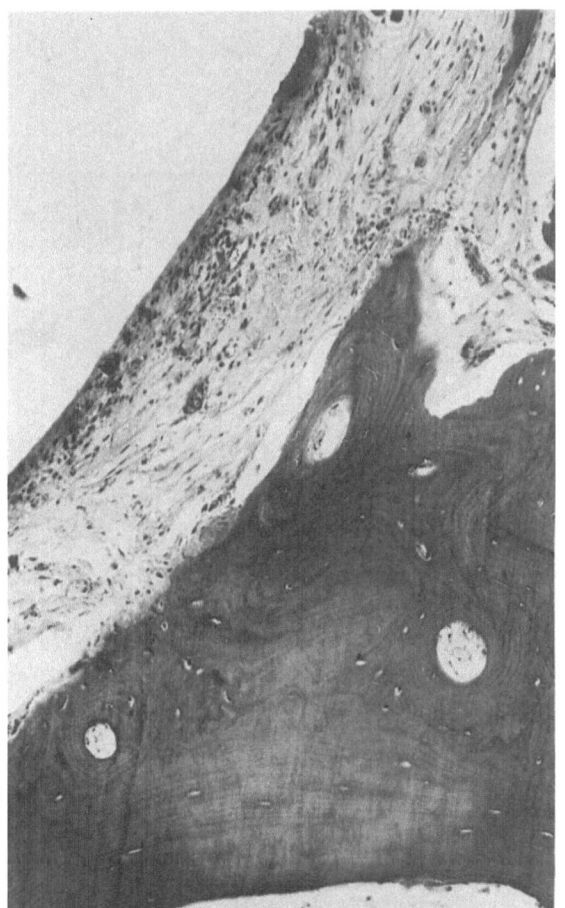

Fig. 22.6. Histological section of the cement interface showing a layer of fibrous tissue between bone and cement.

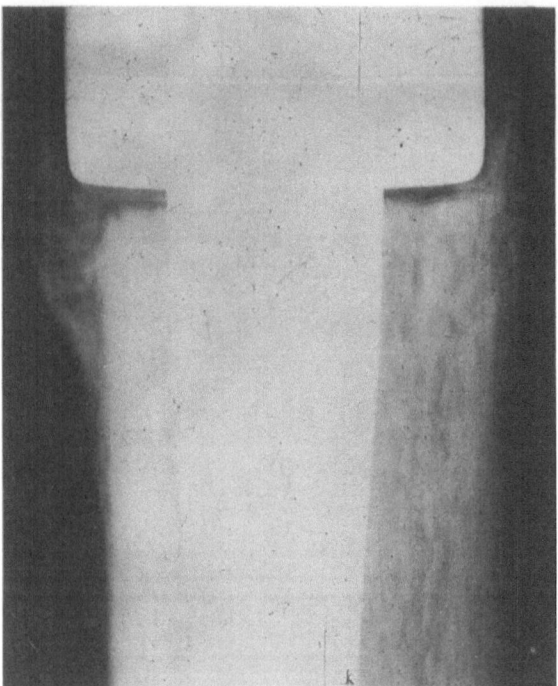

Fig. 22.8. Radiograph of the retrieved tibial component showing the radiolucent space between the shoulder of the prosthesis and bone. Note the pedicle and the difference in bone density between the medial and lateral sides of the tibia.

the left femur. He had a massive mid-shaft femoral replacement but died from pulmonary metastases a year later. Post-operative radiographs and radiographs of the retrieved specimen showed a small radiolucent gap between the shaft of the prosthesis and cortical bone (Fig. 22.11), and penetration of the acrylic cement into cancellous bone. Histological sections showed how the cement had interdigitated with the trabeculae even though some of the trabeculae appeared to be fractured at the cement interface. A broad band of randomly organised fibroblasts was seen at the narrow cement interface and foreign body giant cells lined the beads pro-

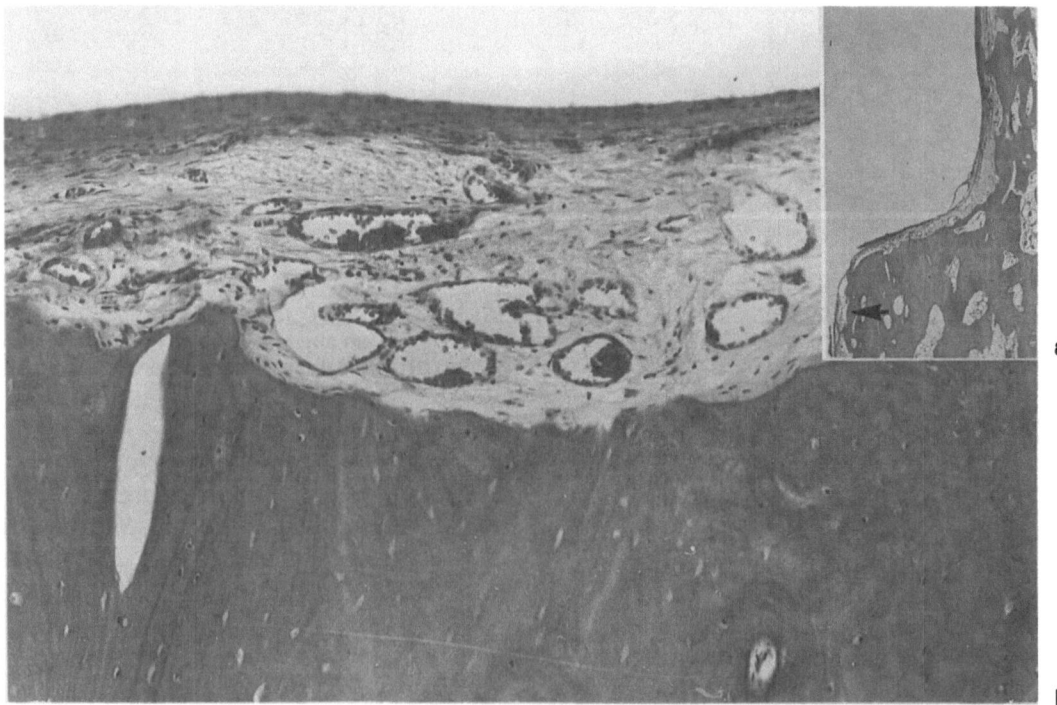

Fig. 22.9. a A low-power histological section of the interface at the shoulder of the prosthesis. The connective tissue interface at the shoulder is continuous with connective tissue at the cement interface (*arrowed*). **b** A higher power view of the connective tissue interface at the shoulder. Note the large number of blood vessels present in this tissue and also the remodelled and necrotic cortical bone below this interface.

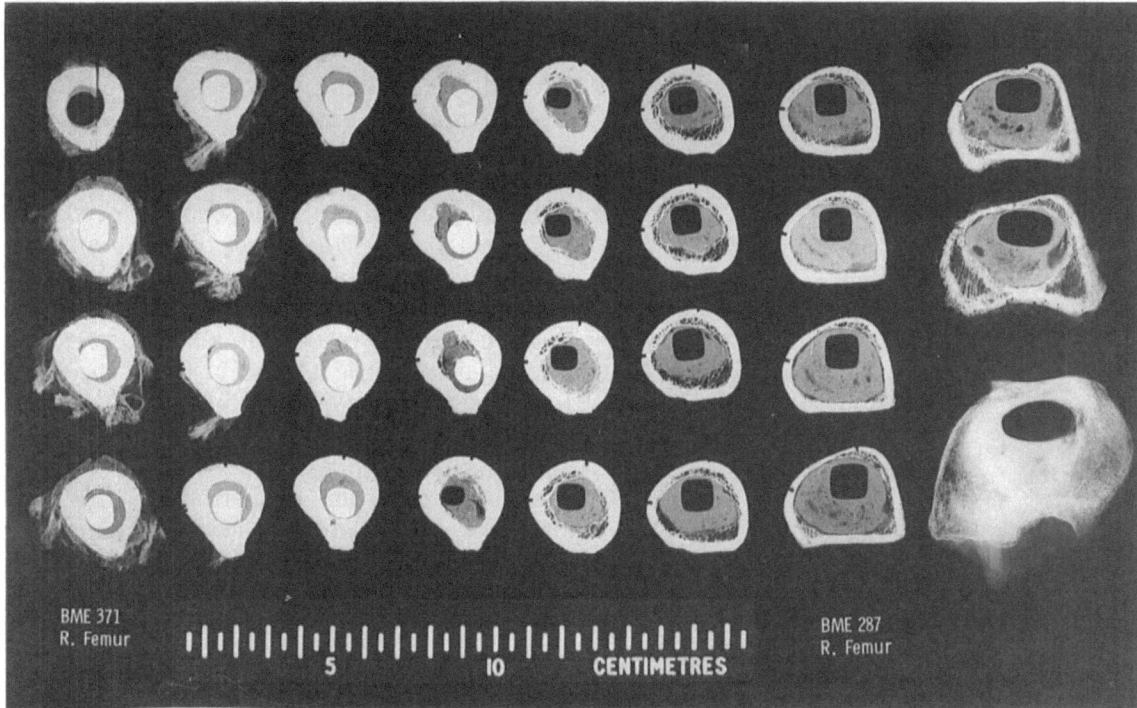

Fig. 22.10. Radiograph of sequential 5-mm thick slices taken through the proximal femoral stem down through to the distal femoral stem. The slice adjacent to the shoulder of the proximal femoral replacement is in the *top left*. The metal has been removed from the slices taken through the distal femoral stem. The slice adjacent to the knee joint is in the *lower right*. Little remodelling has occurred around the proximal stem, whereas a shell of bone has formed around the cement of the distal stem and the inner cortex has become porotic.

jecting from the mass of cement. Some of the trabeculae within the cement were necrotic and surrounded by a thin layer of densely stained cellular tissue (Fig. 22.12). Other trabeculae appeared to be in direct contact with cement (Fig. 22.13).

Case 4

The patient was a boy aged 13 years who developed an osteosarcoma in the distal femur. The tumour was replaced with a Stanmore "growing" prosthesis. Growth of the limb was achieved by preservation of the tibial physis and the use of a femoral component which could be extended by a minor operation. The patient died 6 months after surgery from metastases in the lung and spine.

Sequential radiographs of the retrieved femora from both the operated and unoperated legs showed the vast increase in sub-periosteal bone formation around the

intramedullary stem as seen in other young patients. The inner cortex had become porotic especially on the medial side whilst a shell of bone remained around the acrylic cément (Fig. 22.14).

Case 5

The patient was a 72-year-old man who presented with painful arthrosic disease of the right hip. A McKee—Farrar metal-to-metal prosthesis was inserted which remained trouble-free for 6 years until the patient died from cardiac failure. Radiographs of the retrieved prosthesis showed that a sheath of dense bone had formed around the non-opaque cement at the distal stem tip (Fig. 22.15). Sequential slice radiographs showed that the inner diaphyseal cortex had become porotic. Trabecular struts attached the shell of bone, which had formed around the acrylic cement in the diaphyseal slices, to the outer more dense cortex (Fig. 22.16). A soft tissue inter-

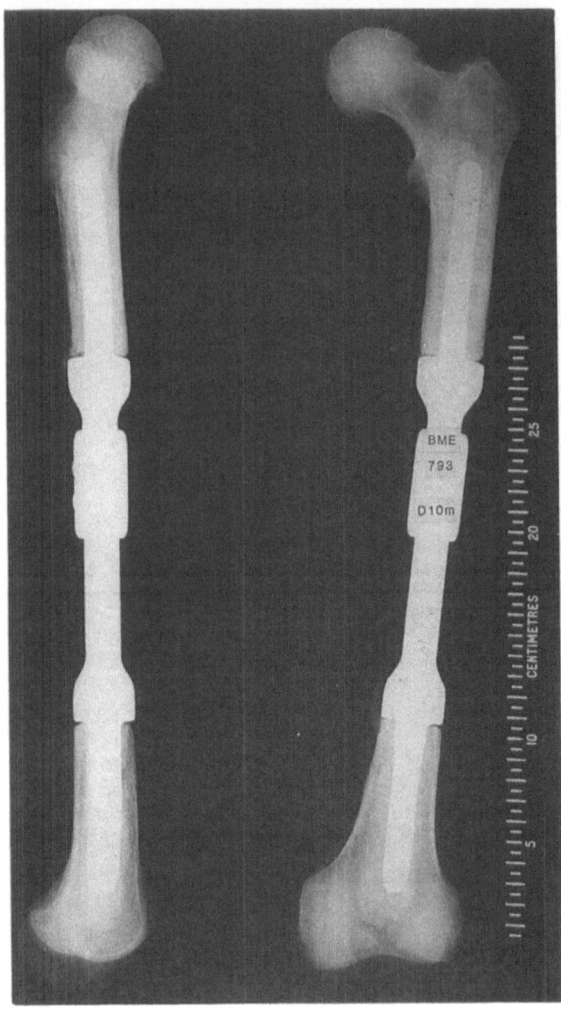

Fig. 22.11. Radiograph of retrieved mid-shaft femoral replacement. Note the penetration of the cement into cancellous bone.

face occurred adjacent to the cement and was composed of macrophages containing small metal wear particles and fibroblasts (Fig. 22.17).

Case 6

A 69-year-old man had an arthrosic hip replaced by a Stanmore prosthesis. The hip gave good service until the patient's death from cardiovascular disease. Radiographs at the level of the lesser trochanter demonstrated that a shell of bone had formed around the cement which was strutted to the outer cortex by trabeculae (Fig. 22.18).

Conclusion

Several remodelling processes were seen around massive endoprostheses. The formation of a pedicle at the transection site can be attributed to load-adaptive bone formation. Formation of this pedicle usually occurs on the medial posterior aspect of the femur, i.e. the side under compressive load due to the combination of the bending moment and the relative stiffness of the metallic prosthesis. In almost every retrieval and in those cases that have been reviewed clinically, a radiolucent space developed between the bone and the shoulder of the prosthesis. This can result in the uncoupling of load transfer from the shoulder of the prosthesis with greater load being transferred towards the stem tip. In Case 1, this resulted in the loss of bone densi-

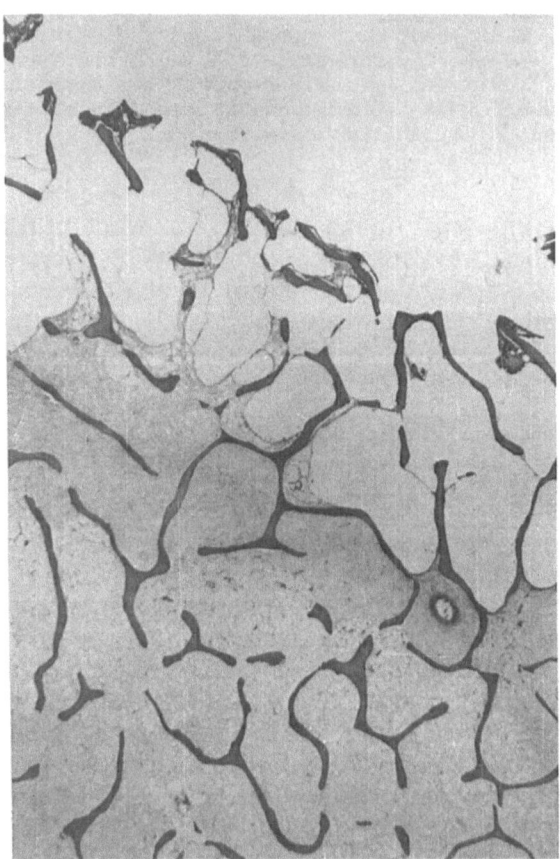

Fig. 22.12. Histological section showing the interdigitation of the cement with trabecular bone. Some of the trabeculae have fractured at the cement interface.

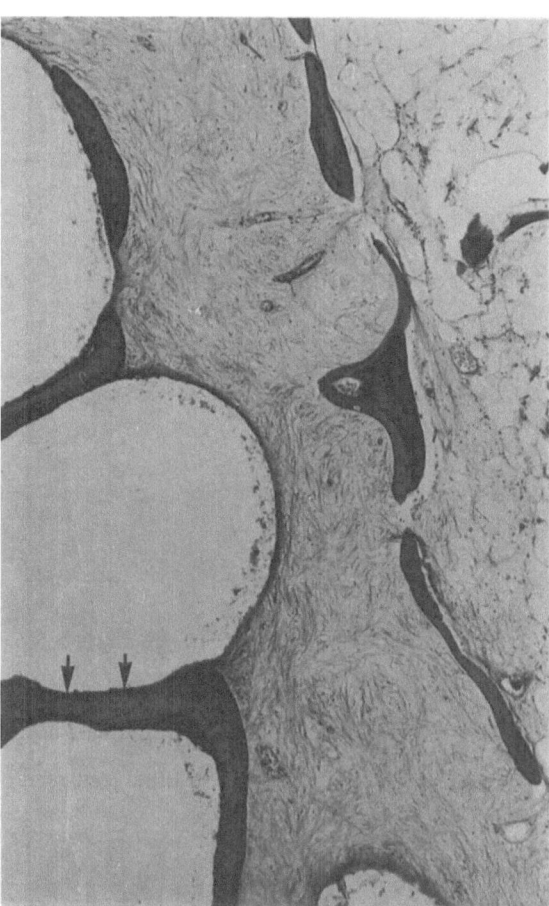

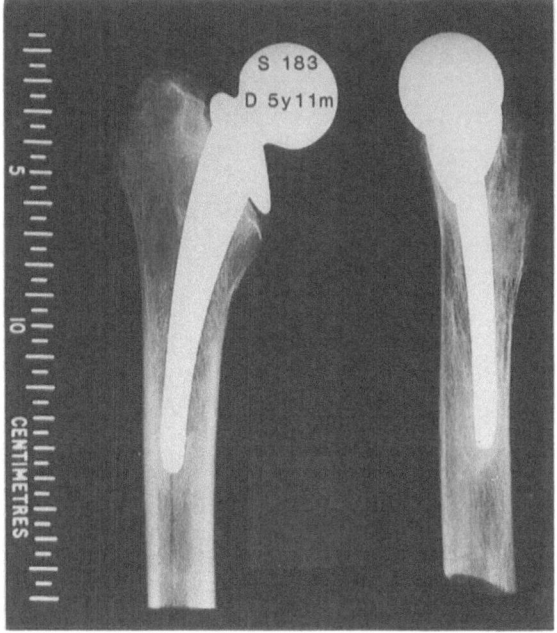

Fig. 22.13. Histological section through the cement interface showing the random orientation of the fibroblastic layer between the cement and marrow and the apparent contact between trabeculae and cement (*arrowed*).

Fig. 22.15. Radiograph of retrieved femoral component of McKee—Farrar total hip replacement.

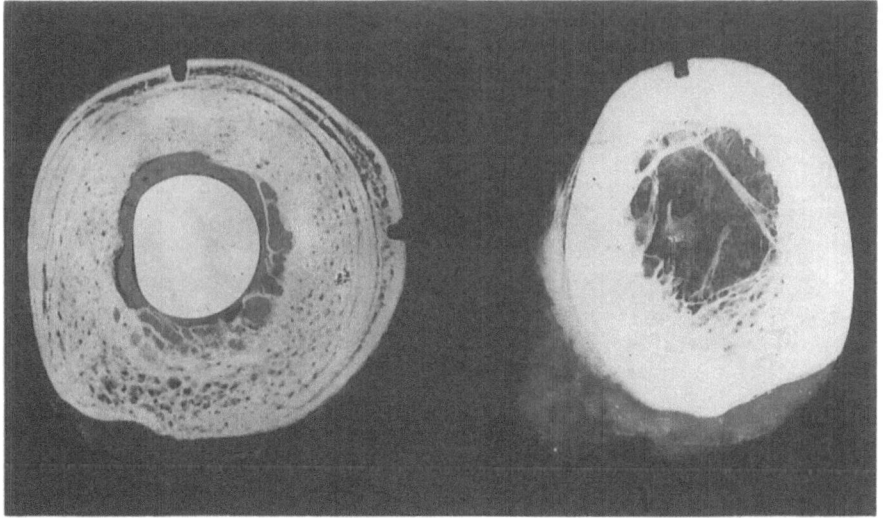

Fig. 22.14. Radiographs of slices taken at the same level from the limb with the intramedullary stem (*left*) and from the untreated limb (*right*).

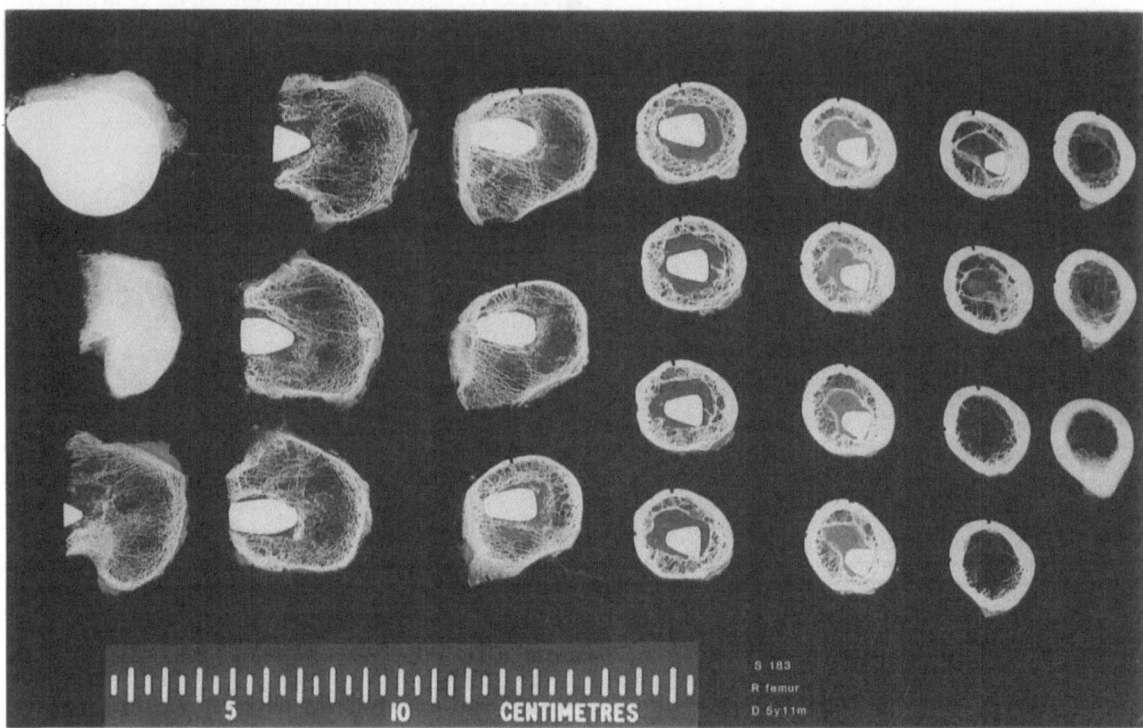

Fig. 22.16. Slice sequence showing the development of a shell of bone around acrylic cement and inner cortical remodelling in the distal slices.

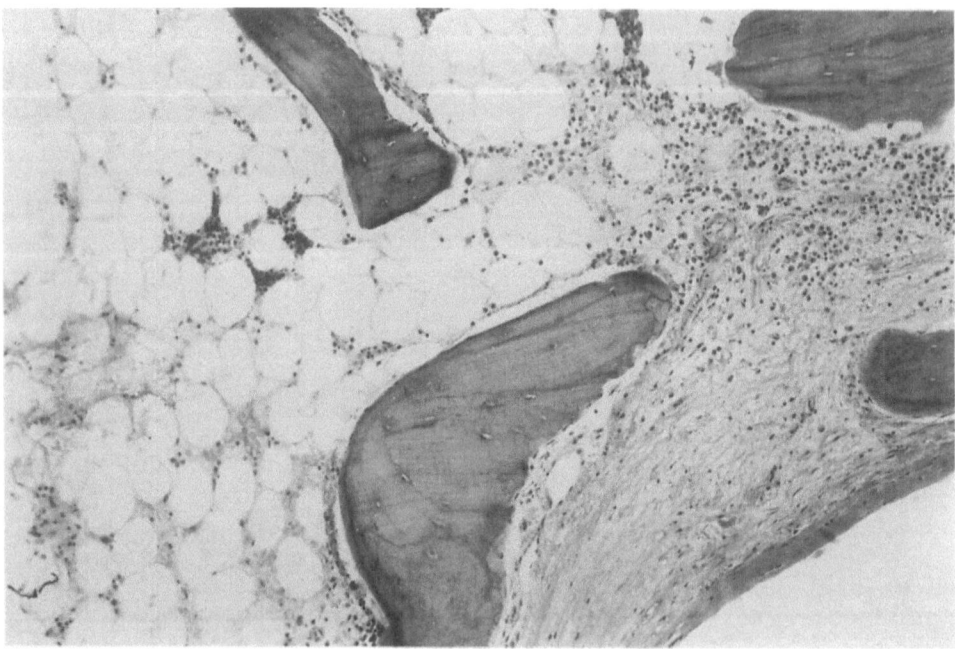

Fig. 22.17. Histological section through the cement interface showing soft tissue adjacent to the cement.

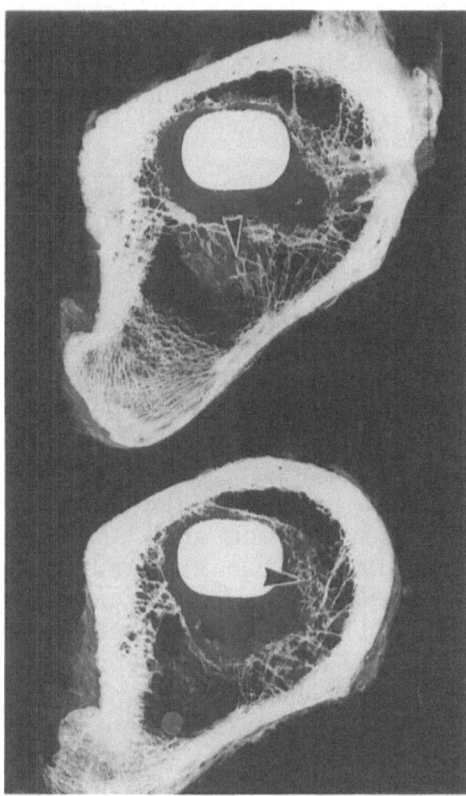

Fig. 22.18. Radiograph of two sequential slices taken at the level of the lesser trochanter showing the shell of bone around the cement. This shell is strutted to the cortex by trabeculae. In places trabeculae appear to have fractured and resorbed (*arrowed*).

ty and the resorption of the pedicle at the transection site. The reason that bone resorbs at the transection site is unclear. However, the fact that the soft tissue present in this space contains numerous blood vessels which invade the cortex suggests that the development of a blood supply to cortical bone is from vessels which enter at the transection site. The large areas of dead cortical bone seen in some of these patients were due to devascularisation of the cortex.

An animal model was used to investigate the effect of femoral reaming and cementation implantation on revascularisation of the canine femur (Rhinelander 1979). Extensive necrosis of the inner diaphyseal cortex was present a year after reaming and introduction of cement. A fibrous membrane was always present in the diaphysis and served as an avenue to the regenerative medullary blood supply. This may be the reason for the soft tis-

sue interface around the cement seen in this study.

Cortical remodelling resulting in a porotic inner cortex was observed in all cases. This is not a result of age-dependent changes for it occurs in young and old patients. It is a feature associated with intramedullary cemented stems. In Case 2, the femur had remodelled around the stem of the femoral component which had been in place for 18 months, but there was no remodelling of proximal bone. This feature also occurs around femoral components of total hip replacements but is associated only with the diaphyseal cortex.

The shell of bone which develops around the cement, usually in the diaphyseal cortex, appears to form as a result of load adaptation. Working on sheep and canine models, it seems that response to the hoop stresses generated by pistoning of the implant is the cause of the bone shell (Brown et al. 1988). If this is so, there must be some movement of the implant relative to the femur. It was evident that resorption and fracture of the trabeculae supporting the shell of bone had occurred. This could have been due to overload of the trabeculae. Although it did not lead to failure of the prosthesis in this case, it could be envisaged that further fracture may have led to loss of support proximally and failure of the prosthesis.

Despite the remodelling that occurs around intramedullary stems of both massive and standard prostheses, cement fixation has proven over many years to be a reliable and effective method of fixation.

Acknowledgement. Dr Blunn, who presented this paper, is indebted to his co-author for her collaborative work.

References and Further Reading

Brown TD, Peterson DR, Radin EL, Rose RM (1988) Global mechanical consequences of reduced cement/bone coupling rigidity in proximal femoral arthroplasty. A three dimensional finite element analysis. Biomechanics 21:115—129

Rhinelander FW, Nelson CL, Stewart RD, Stewart CL (1979) Experimental reaming of the proximal femur and acrylic cement implantation — vascular and histologic effects. Clin Orthop 141:75—89

Discussion

The Chairman - **Mr Ling**

The Panel - **Professor Fitzgerald**
         - **Professor Howie**
         - **Dr Blunn**

Mr Gruebel-Lee: Professor Howie, I am involved in a long-term prospective study comparing slight clinical differences in total hip replacement. I chose to change the shape and length of stem, but you cannot change the shape and use polymethylmethacrylate unless you use a long stem. I am using a curved stem 17 cm instead of 11 cm in length. I have, therefore, two variables which I cannot reduce. Your two variables are cement and non-cement, and a curved stem. Are we unwise to choose two variables that will make analysis difficult?

Professor Howie: Yes, I think you may be right, but our variables seemed to be satisfactory in the USA so we thought we would give them a proper trial. We did an audit for the Ethics Committee's approval to decide whether we were justified ethically in going on.

Dr Lee: I would like to ask Professor Fitzgerald about the design of his collarless implant. You took the HG2, which, of course, has a fairly massive collar. As I understand it, the manufacturers removed the collar from that stem for you. I did not see you show a picture of the two to compare them directly, but looking at one or two X-rays it seemed quite clear to me that there was still some collar left on the implant. In addition, there was a satin finish to the stem with dimples on the side. I am rather worried, therefore, about your "collarless" description.

Secondly, with regard to the results that you presented on loss of medial bone height, you said that you lost about 2 mm with your collared and 7 mm with the collarless version. This is totally and absolutely the opposite of our experience. Again, you did not show any X-rays to indicate, in particular, whether or not there was any cement around the surface of the neck of the medial calcar in either the collarless or the collared implants.

I would like your comments on these two points: the first is your justification in calling your modified HD2 a collarless stem; the second is how you assess the difference between the height loss in the medial calcar region.

Professor Fitzgerald: Does throwing load on to the calcar have an impact? That was the question. Whether we call the manufacturer's version of the HD2 without a collar "collarless" or something else is a matter of semantics as far as I am concerned. In no way did we anticipate that the device without a collar would subside. It is designed to do the opposite. Anybody who has used the HD2 knows that you cannot achieve calcar/collar contact for a significant amount of the time. So what does that mean? In this small but controlled series with one variable, I can only give you our numbers as far as loss is concerned. We saw appreciable loss when we unloaded the cut end of the femoral neck which was not loaded with a collar. There was no cement over it since we were careful not to overlap the cement. I am sure there was some cement contact in some of the collared prostheses because you can only achieve point contact with the collar sometimes, especially on the HD2 as it is designed today. It is very difficult to give the percentages but in some patients cement was caught under the collar on to the calcar-cut surface. Nevertheless, when you are able to load the collar with this particular device, you are able to protect the proximal medial cement. We have watched it now over a 5-year period.

Dr Lee: Could I follow that up a little? You said that you managed to get optimal contact in only 47% of your collared cases. Did you notice any difference in behaviour in that approximate 50/50 split? Secondly, as soon as there was loss in height you lost contact and the ability to transmit load by the collar. Presumably your collared implant then, in effect, becomes collarless, and yet the results are still significantly different.

Professor Fitzgerald: Yes, when we looked at the 41 patients in that group, half had what I considered to be acceptable calcar/collar contact. The device behaved as a collarless device when it did not have contact with the cut surface. With resorption of the cut surface of the calcar area, the bone will be unloaded and with this particular device more force will be transferred into the cement.

Time will tell, but I would anticipate a higher failure rate in that group in the future. The message is clear that if you use a collar device, you had best do what is necessary surgically to achieve good contact. If you do not, you put that patient at risk of early failure.

Professor Howie: When a total knee is radiographed, a 1-mm lucency will be missed if you are 3° off alignment. Could there be a similar error with the hip because you have a deceptive collar resorption? Is it worth while trying to use a fluoroscope?

Professor Fitzgerald: They are all fluoroscoped now.

Professor Howie: Are those results based on the fluoroscoped ones?

Professor Fitzgerald: Some are and some are not at this stage. We rotated the specimen but interestingly the lines are not so well hidden in the hip as in the knee. You may hide the size of the line but not its existence.

Dr Lee: Professor Howie is obviously someone who has no absolutely fixed view about whether an implant should be cemented or not. He would clearly be happy to do either. Therefore, the simple question is, what is his overall view about whether one should or should not use cement?

Professor Howie: We are still doing a lot of cemented hips especially if the life expectancy of the patient is less than 10 years. We did a cementless trial because of the long-term problem of revision. I think that revising a cementless porous ingrown prosthesis may probably be harder than revising a cemented one, but we may be wrong.

Mr Gie: Professor Howie, I think we have been considerably influenced by the difficulty of revisions using the curved stems with cement going well down the canal. Now, with straight canals and cement only 1 cm beyond the tip, there is no doubt that cemented revision has become a far easier prospect. I would say that the situation has almost reversed itself. Would you agree with that?

Professor Howie: Yes, I would.

Mr Macdonald: Professor Fitzgerald, you stated that you did not see any cement fracture. What sort of cement fracture were you looking for and how did you miss it?

Professor Fitzgerald: You find the cement starts to fracture in the collarless device.

Dr Lee: Is that on the medial side underneath the calcar area?

Professor Fitzgerald: Correct.

Dr Lee: The implant that you were using was a very stiff, quite massive implant whether it had a collar or not. I cannot understand the mechanism by which you came to crack or shatter bone cement. If you are going to shatter bone cement a considerable strain must be put on it quite quickly, and unless your fixation was severely deficient in the distal portion, I cannot see how the implant could twist and bend to such an extent that there was cement fracture. What is the mechanism for fracture of your cement?

Professor Fitzgerald: What happens is that you lose the bone. You have lost your external support.

Dr Lee: You have still got to strain it.

Professor Fitzgerald: It gets strained. I can just tell you what we found on the radiographs. The stiff stem is designed not to move.

Chapter 23

Hydroxyapatite Crystals in Bone Ingrowth and Comparison of Autografts and Allographs: Preliminary Report

R.H. Fitzgerald Jr

A study was made to assess the efficacy of tri-calcium phosphate hydroxyapatite crystals in a primary canine arthroplasty using a cement-less device with voids between it and its ingrowth pad or surface, and the bone. This led to a second study which evaluated the superiority of autograft over allograft as a concept. We looked at implanting an ingrowth arthroplasty without bone graft, with allograft and then with autograft in a canine model with failed cemented arthro-plasty.

Study I

This study looked at the efficacy of using tri-calcium phosphate hydroxyapatite particles to enhance the biological fixation of a canine titanium fibre metal femoral component with a non-interference fit.

Methods

The device chosen was a titanium canine femoral component. The acetabular compo-nent was cemented for cost reasons. In the tricalcium phosphate model, group 1 had an interference fit, group 2 had voids and the voids were filled with tricalcium phosphate hydroxyapatite crystals in group 3.

The specimens were collected, X-rayed and then divided so that alternate adjacent sec-tions could be prepared for either biomechan-ical analysis, push-out testing or histological and biomechanical testing. The push-out test-ing was performed in the Biomechanical Laboratory using appropriate testing tech-niques. We also looked radiographically at subsidence resorption in movement of the prosthetic devices. A point counting tech-nique was used to assess the number of ingrowth areas initially and then how many areas were filled with fibrous, fibro-cartilagi-nous or bony tissue. The adjacent area just outside the pad and the space occupied by fibrous tissue, osseous tissue or tricalcium phosphate hydroxyapatite crystals was then examined in the first primary study, and dead bone versus live bone in the revision study.

There were originally 25 animals in the pri-mary study with tricalcium phosphate hydroxyapatite crystals, but 6 were lost, one to dislocation, one to a volvulus and one in group 3 to infection. All the implants were inserted in a laboratory operating theatre equipped with ultraviolet light.

Results

Three fluorochrome labels were used to look at bone activity. In group 1, all three labels could be identified over the 12-week period. In groups 2 and 3, all three fluorochrome labels could only be found at the tip in level 7. The fluorochrome labels were identified at all levels at 6 and 9 weeks in groups 2 and 3. That first label could only be found, however, in the animals with tricalcium phosphate at the tip.

The proximal, middle and distal aspects of the femoral component were examined at the mechanical testing, and the differences in shear strength between group 2 at the different levels and between groups 1 and 3 could be seen.

The percentage of the fibre metal pad voids filled with bone was consistent between the proximal, middle and distal aspects of the pad in group 1. However, in groups 2 and 3 the voids were less filled with bone in the proximal and middle aspects of the thermal component compared with the distal aspect. The differences in the proximal aspects between groups 2 and 3 versus group 1 and in the middle were statistically significant.

Conclusions

We concluded from these data that the hydroxyapatite tricalcium phosphate crystals did not enhance the biological fixation of a canine component with a non-interference fit, nor were they effective for ingrowth of the femoral component in patients with a non-interference fit. We felt that the crystals were neither osteo-conductive nor osteo-inductive. If anything, they appeared, in our opinion, to occupy space in the area adjacent to the pads and preclude osseous ingrowth.

Study II

This study compared autograft and allograft in the enhancement of biological fixation of a canine porous femoral component.

Methods

There were three groups in the second study, including the control group, in which the loosened cemented arthroplasty was removed and an ingrowth femoral component was inserted. Three sizes were available to us and we used the best fit. In group 2, six dogs received a fresh iliac crest graft, while in group 3, six dogs received frozen allograft. All three groups had their iliac crest exposed to control for surgical intervention. The animals were allowed free weight bearing. They had a continuous tetracycline label and radiographs were taken monthly. The animals were sacrificed 12 weeks after the revision arthroplasty. One dog with an allograft bone was lost due to deep infection and there was one control animal that had retained cement that we did not see on inter-operative radiographs or at surgery. We felt that it altered the stability of our prosthesis so we removed that dog from further analysis. The specimens were collected, X-rayed and then subjected to histological and biomechanical evaluation as in Study I.

Results

Subsidence averaged 5 mm in the control animals, 0.8 mm in the autograft group and 2.4 mm in the allograft group.

Histologically, again looking at proximal, middle and distal aspects of the femoral components, the mean percentage ingrowth was almost nil in the control animal, 22% in the autograft group and 17% in the allograft group. The difference between control and autograft and control and allograft was statistically significant ($p < 0.05$).

It appeared that ingrowth occurred first at the edge of the pad in the primary and the revision system, and was only apparent later in the middle of the pad. The middle and distal sections showed ingrowth in 7% of the control animals, 11% of the autograft group and 11% of the allograft group. The difference in ultimate strength between control and autograft and autograft versus allograft was statistically significant ($p < 0.05$). Energy to failure was not statistically significant ($p > 0.05$). Differences that have been noted throughout were apparent in the final histological ingrowth.

There appeared to be a relationship between stability as-judged by lucent lines and subsidence, and between those factors and ultimate strength. Ultimate strength was statistically significant ($p<0.05$). There was also a statistically significant difference between those animals with and those without radiolucent lines which was not surprising. Again, similar information was apparent with subsidence or its absence related to mechanical properties.

Posterior tightness was noticed in the animal model. As the prosthetic device was put into the femur, it was very hard to insert bone graft in the proximal aspect. It was easy to insert the graft in the middle and distal sections posteriorly but very hard proximally.

Conclusion

In summary, we felt that, histologically, bone ingrowth was greatest proximally in the grafted groups and anteriorly in all groups. Biomechanical strength was greater proximally in the grafted animals and this correlated with bony ingrowth. The biomechanical performance related directly to the amount of ingrowth. The autograft animals fared better than the allograft group in achieving implant stability. At 12 weeks, however, the incorporation of the ingrowth device in those animals receiving autograft or allograft appeared equal, and implant stability correlated well with bone ingrowth in biomechanical strength.

Chapter 24

Fluoroscopically Controlled Radiographs in Examination of the Bone—Implant Interface: Preliminary Results

D.W. Howie, C.M. Steele-Scott and D. Waters

The purpose of this study was to examine the cementless bone interface around porous-coated total knee arthroplasties. A radiographic technique was used to define the early changes, particularly under the tibial tray. While bone ingrowth will occur around porous-coated implants, the amount of ingrowth appears to be small (Cook et al. 1989). The tibial tray in particular will rock under the load, resulting in significant micromotion at the bone—implant interface. It may not be realistic, therefore, to expect significant bone ingrowth to occur and maintain fixation.

Methods

There are two methods commonly used to examine the bone—implant interface. The first is to rely on examination of retrieved specimens. Failed implants, however, may tell us little about what is occurring around a successful implant. Although postmortem specimens may be useful, the clinical status of these arthroplasties is often unknown.

At the beginning of this study our intention was to insert cementless total knee arthroplasties sufficiently well that they would not need revision within the first 10 years. So we did not expect to have the opportunity to examine revision specimens from our own series within this period. The arthroplasties were, in general, performed in patients with a 10-year life expectancy and so postmortem specimens were also unlikely to be obtained.

The other method commonly used to analyse the bone—implant interface is radiography. There are, however, significant problems with the interpretation of plain radiographs, largely due to projectional errors. An image intensifier was used in this study to allow orientation of the radiographic beam parallel to the bone—implant interface. The postoperative changes at the interface were examined in the first year.

Twenty-nine cementless PCA total knee arthroplasties were examined. The PCA knee scores were used to give a clinical perspective to this study and long leg radiographs were used to determine the post-operative alignment of the knees.

Fluoroscopically Controlled Radiographs

We have compared fluoroscopically guided radiographs and plain radiography (Mintz et al., in press). A PCA tibial tray was inserted into cadaveric specimens. The tibial tray was separated from the bone by connective tissue taken at revision surgery from the bone—implant interface. A series of fluoroscopically controlled and plain radiographs was taken to detect radiolucent lines and, if present, to measure their thickness.

We examined the thickness of the radiolucencies on the AP and lateral projections. Ten separate series of plain and fluoroscopic radiographs were performed. The fluoroscopically controlled radiographs detected radiolucencies in all zones in the AP and lateral projections while plain radiographs rarely did so.

Results

The clinical results in this group of patients at 1 year were either excellent or good except for two cases with anterior knee pain. No revisions were undertaken.

Preliminary assessment of the fluoroscopic radiographs demonstrated an increase in the number of radiolucencies within the first 6 months compared to radiographs taken immediately after surgery. The lucencies were commonly seen beneath the tibial tray and adjacent to the anterior and posterior flanges of the femoral components. The incidence of radiolucencies appeared to increase slightly between 6 months and 1 year. The number of loose beads beneath the tibial component increased between the initial post-operative radiographs and those taken at 1 year.

Conclusions

An accurate image of the bone—implant interface around cementless total knee arthroplasties can only be obtained using fluoroscopically controlled radiographs. Preliminary assessment of the use of these radiographs in monitoring the appearances around porous-coated total knee arthroplasties suggests there is an increase in the incidence of radiolucencies in the first 6 months after surgery. Longer term review will be required

to determine whether the radiolucencies are progressive. It seems likely that macro-interlock, in conjunction with connective tissue ingrowth, is the major contributor to fixation of these implants.

Acknowledgement. Professor Howie, who presented this paper, is indebted to his co-authors for their collaborative work.

References and Further Reading

Cook SC, Barrack RL, Thomas KA, Haddad RJ (1989) Quantitative histologic analysis of tissue growth into porous total knee components. J Arthroplasty (Suppl) 4:533—543

Mintz AD, Pilkington AJ, Howie DW (in press) Comparison of plain and fluoroscopically guided radiographs in assessment of knee arthroplasties. J Bone Joint Surg (Br)

Discussion

The Chairman - **Mr Ling**

The Panel - **Professor Howie**
          - **Professor Fitzgerald**

Mr MacDonald: Looking at the results on the tibial trays, is it fibrous fixation or are the trays sitting on the pegs, and that is the site of major loading?

Professor Howie: It seems that we get bone ingrowth in retrieval specimens; we saw areas of lucency around the tibial trays as well. If you can see a fair proportion of lucencies, there are probably more. I am sure that you can find isolated areas of bone ingrowth around the pegs, but I still think that there is probably considerable connective tissue supporting the implant.

Mr Smith: You did not mention the patella. How often do you see radiolucent lines around the patella implant, because these metal-backed implants have been shown to have a failure load which may occur early?

Professor Howie: We do not have a good method of imaging the patella implant. We take skyline views and also fluorochromes. We do not think we would have seen any lucencies if we had just used plain X-rays.

Mr Ling: In the revision model that Professor Fitzgerald investigated, the autografts were better than allografts in achieving the stability that was important to ingrowth. In the longer follow-up was there any difference between them?

Professor Fitzgerald: We found consistently better ingrowth with the autograft than with the allograft but it was not at a statistically significant level.

Mr Ling: Was it the same for the mechanical testing of the specimen?

Professor Fitzgerald: Yes, except for one.

Professor Howie: Professor Fitzgerald, is it correct that fluorochromes show bone ingrowth?

Professor Fitzgerald: Fluorochromes do show ingrowth. When we used a serial fluorochrome label in the primary model, the animal with an interference fit picked up all three labels. The animals with a void or with tricalcium phosphate only picked up the last two labels. Ingrowth in those animals therefore occurred later.

Professor Howie: How can you be sure that it is live bone that you are seeing in the revision model?

Professor Fitzgerald: The fluorochromes tell you bone is live, and measurement of the thickness of the label confirms when it occurs.

Professor Howie: It has been said that fluorochromes will label dead bone as well. Do you have any problems with that?

Professor Fitzgerald: We did not have any problems. Bone will pick up antibiotics, and antibiotics will diffuse into dead bone. When you look at a sequestered piece of bone you may pick up a little fluorochrome label but you will not have a demonstrable band.

Mr Smith: What size were the hydroxyapatite crystals? Is the size anything to do with the amount of bone induction and osteo-conduction that occurs? Have you looked at smaller crystals or different shapes?

Professor Fitzgerald: It is expensive enough to look at one size. To put 25 animals through this investigation was an expensive and time-consuming endeavour. I cannot remember the crystal size but it is exactly identical in size and range to that which is used for the prospective human studies.

Mr Griffiths: Dead bone, of course, will take up the fluorochrome, and will be a pattern which you may not be able to distinguish, but how long after the ministration did you look at it?

Professor Fitzgerald: The animals that had allograft, and therefore a significant amount of dead bone, received a continuous label for 12 weeks, so we obtained a thick band. We had no trouble in picking up and identifying our peripheral allograft which had not been incorporated.

Chapter 25

Conclusion

R.A. Elson

At the end of this remarkable meeting, it was my privilege to have been asked to summarise the events of the two stimulating days of presentation by and discussion with a distinguished group of international authorities. John Older not only succeeded in bringing the group together but has also persuaded the Faculty to allow this written account of their views to be published, so making them available to a wider audience. I believe that the book affords a unique record of current opinion as to the nature and behaviour of the bone—cement interface and its relationship to the uncemented prosthetic environment.

I had expected to leave Midhurst with a concise concept of how to choose my next implant and by what method to insert it into the patient; far from it! The audience then, and now the reader, cannot expect a conclusion at this stage of development. Every factor needs 10 years for evaluation and integration into the whole spectrum of possibilities for improvement which now face us. It is true that some concepts are slowly consolidating, but even within a broad understanding of the cemented hip arthroplasty (apparently not greatly altered since the early days of Sir John Charnley), there exists a continuous stream of new pieces of information which pushes us from side to side. We hope that by improved audit, one day it will be possible to obtain reliable results from a bigger population of surgeons dedicated to submitting reliable data.

One thing is certain. While we recognise the heightened interest in the uncemented device with anchor by porous ingrowth, osseous integration or bulk interference, we must also recognise that the value and potential of acrylic cement is far from exhausted.

I suppose that I have to admit a certain satisfaction in feeling able to make this statement. At the same time, I am wistful and acutely conscious of the need for the next generation to exploit and to evaluate truthfully all current minutiae of improved cementation and implant design. This acrylic cement, a remarkable material, may afford a unique barrier to the erosive action of wear products from within the joint; the containment of these may be the final arbiter determining long-term survival. Properly and safely intruded, and while perhaps exploiting the time-dependent properties of the material, the coupling of the bone to the more rigid metal or ceramic or to a more compliant plastic cement may continue to be exploited in the manner of which I cannot conceive an uncemented device.

At the meeting I saw fit to contrast some of the opinions expressed, but in this book comment is inappropriate. The reader of these accounts will already have a special interest in the bone—cement interface. Certainly his interest will be enhanced after reading.

As one of Sir John Charnley's pupils, I cannot fail to speculate what his comments would have been at the close of this remarkable series of presentations. There might be a tendency to say "I told you so", but we should resist such temptation and be prepared for change of mind. Charnley was criticised for doing this repeatedly during the evolution of his hip arthroplasty. On the contrary, he should be admired for not only accepting the need for change of mind, but then for meticulous evaluation of the outcome — and then perhaps changing it again! One other certainty is that Sir John would have been delighted to have seen the meeting taking place at Midhurst, a Hospital of which he was so fond.

Subject Index